COMPETITION, REGULATION, AND CONVERGENCE

*Current Trends in
Telecommunications
Policy Research*

TELECOMMUNICATIONS

A Series of Volumes Edited
by Christopher H. Sterling

COMPETITION, REGULATION, AND CONVERGENCE

CURRENT TRENDS IN TELECOMMUNICATIONS POLICY RESEARCH

Edited by

Sharon Eisner Gillett
Massachusetts Institute of Technology

Ingo Vogelsang
Boston University

LAWRENCE ERLBAUM ASSOCIATES, PUBLISHERS
Mahwah, New Jersey   London

The final camera copy for this work was prepared by the author, and therefore the publisher takes no responsibility for consistency or correctness of typographical style. However, this arrangement helps to make publication of this kind of scholarship possible.

Lawrence Erlbaum Associates, Inc., Publishers
10 Industrial Avenue
Mahwah, New Jersey 07430

Cover design and typesetting by Spot Color, Inc., South Riding, Virginia

Library of Congress Cataloging-in-Publication Data

Telecommunications Policy Research Conference (26th: 1998)
    Competition, Regulation, and Convergence: Current Trends in
    Telecommunications Policy Research / edited by
    Sharon Eisner Gillett, Ingo Vogelsang.
            p.   cm.
        Includes bibliographical references and index.
        ISBN 0-8058-3484-2 (cl. : alk paper)
        1. Telecommunication policy—Congresses. 2. Competition—Congresses.
        I. Gillett, Sharon Eisner. II. Vogelsang, Ingo. III. Title.
        HE7645.T45   1998
        384—dc21   99-37900
                                                       CIP

Books published by Lawrence Erlbaum Associates are printed on acid-free paper, and their bindings are chosen for strength and durability.

Printed in the United States of America
10 9 8 7 6 5 4 3 2 1

Contents

Authors

Debra J. Aron
LECG, Inc.

Timothy J. Brennan
University of Maryland, Baltimore

Arturo Briceño
OSIPTEL, Peru

David D. Clark
Massachusetts Institute of Technology

Thomas A. Downes
Tufts University

Scott C. Forbes
Pennsylvania State University

Daniel Fryxell
Carnegie Mellon Univesity

David Gabel
Queens College

Sharon Eisner Gillett
Massachusetts Institute of Technology

Shane M. Greenstein
Northwestern University

Heather E. Hudson
University of San Francisco

Junseok Hwang
University of Pittsburgh

Scott K. Kennedy
Independent Analyst, Florence, MA

Thomas Kiessling
Harvard University

William Lehr
*Massachusetts Institute of Technology
and Columbia University*

Lawrence Lessig
Harvard Law School

Viktor Mayer-Schönberger
*University of Vienna, Austria,
and Harvard University*

Paul Milgrom
Stanford University

Bridger M. Mitchell
Charles River Associates

Milton Mueller
Syracuse University

Paul Resnick
University of Michigan

Jorge Reina Schement
Pennsylvania State University

Marvin Sirbu
Carnegie Mellon University

P. Srinagesh
Charles River Associates

Ingo Vogelsang
Boston University

Kanchana Wanichkorn
Carnegie Mellon University

Jonathan Weinberg
Wayne State University

Martin B. H. Weiss
University of Pittsburgh

Steven S. Wildman
Northwestern University and LECG, Inc.

Acknowledgments

The Telecommunications Policy Research Conference (TPRC) has now run for 26 years. Throughout this history, many of the papers presented have resulted in journal articles and other publications. For most of the years, there was not a regular proceedings publication.

In 1994, the combined efforts of Christopher H. Sterling, the series editor of the Lawrence Erlbaum Associates Telecommunications Series, Hollis Heimbouch of LEA, and TPRC Board Members John Haring, Bridger Mitchell, and Jerry Brock, resulted in a new arrangement for LEA to publish a series of selected papers from the annual TPRC. This is the fifth volume in the LEA series, and it contains papers selected from the 26th annual conference, held in October 1998.

We greatly appreciate the assistance we received in dealing with the difficult task of selecting papers for this volume. We consulted with the TPRC program committee, session chairs and discussants, and various others who volunteered their time to referee. There were over 70 papers presented at the conference, and there were many outstanding contributions that we regret we could not include because of space limitations.

We have worked to fulfill our promise to deliver a published text in time for the 27th annual conference. We could not have done this without expert production work. We thank: Teresa Horton for copyediting (in softcopy for the first time in the history of this volume); Jennifer Sterling and her staff for typesetting; and Linda Bathgate and her colleagues at LEA for shepherding this book through all the phases of production.

We also owe special thanks to: Walter Baer, Chairman of the TPRC Board of Directors, who brought us together to edit this volume; Jeff Mackie-Mason and David Waterman, editors of the 1997 proceedings volume and members of the TPRC board, who gave generously of the wisdom borne of their experience and assisted us at several critical junctures; and Merrick Lex Berman, administrative assistant for the MIT Internet and Telecoms Convergence Consortium, who helped us out in countless ways.

Finally, neither the 1998 conference nor this volume would have been possible without the generous financial support of TPRC's sponsors. In particular, the long running grant support of the John R. and Mary Markle Foundation deserves special mention. In addition, the 26th TPRC had the benefit of a matching grant from the W.K. Kellogg Foundation, as well as donations from Telecom Italia Mobile, America Online, Inc., Ameritech, Bell Atlantic, BellSouth, CableLabs, Cablevision, MCI, Oracle Corp., Sprint, US WEST, the Benton Foundation, National Economic Research Associates, LECG, Inc., Telecommunications Research Inc. (a division of Nathan Associates Inc.), the Foundation for Rural Service, Strategic Policy Research, Charles River Associates, Darby Associates, and HAI Consulting, Inc.

Sharon Eisner Gillett
Ingo Vogelsang

Preface

The Telecommunications Policy Research Conference (TPRC) holds an unrivaled place at the center of national public policy discourse on issues in communications and information. Yet ironically, TPRC's uniqueness arises not from sharp focus but from the inclusive breadth of its subject matter and attendees. Unlike other conferences, participants in TPRC do not come exclusively from academia, industry, or government, but rather from a mix of all three. Indeed, TPRC is one of the few places where multidisciplinary discussions take place as the norm. If TPRC were soup, economists would be the stock (the largest single ingredient and the fundamental flavor), with lawyers, social scientists, MBAs, engineers, and others adding variety and spice to the mix. Of course, the result in either case is a tasty blend.

The breadth of TPRC also extends to its definition of telecommunications. The general population may equate telecommunications with telephony—one thing you can always count on at TPRC is a discussion of whether the conference should change its name! But at least in the context of TPRC, telecommunications policy has come to encompass the many business, social, and regulatory issues raised by the provision and use of a wide variety of communications and information services. In addition to telephony, these services include the broadcast mass media and, most recently, the data communications and information services enabled by the Internet.

This volume represents the 26th annual TPRC gathering and reflects a matrix of the topics considered by the attendees. The papers collected in this year's proceedings under the themes "Competition, Regulation, Universal Service, and Convergence" provide an exciting sample of the approximately 80 conference papers selected (from more than twice that number submitted) for presentation.[1]

---

1. The entire program of the conference with links to papers and authors is available at the TPRC website, http://www.si.umich.edu/~prie/tprc.

In assembling the program for the 1998 TPRC, we found that the range of interests and the needs of the conference attendees demanded greater emphasis than ever on issues associated with the rapid diffusion of the Internet: privacy, security, equity, access, reliability, economic stability, governance, and technological structures. In convening a conference to tackle such disparate topics, we relied on a committee of leaders from across the disciplines contributing to the conference. The 1998 Program Committee included Ira Barron from US West; Erik Brynjolfsson and Sharon Eisner Gillett, both from the Massachusetts Institute of Technology; Robert Horwitz from the University of California at San Diego; Jonathan Levy from the Federal Communications Commission; Jessica Litman from Wayne State University's School of Law; Luigi Prosperetti from the University of Milan; and Padmanabhan Srinagesh from Charles River Associates. The quality of the papers in this volume is only one mark of the hard work and valued advice they provided during the year of planning for the October gathering.

As co-chairs of the 26th conference we respectfully join the company of our distinguished colleagues who have chaired the previous 25 conferences. Knowing that several hundred very smart people will come together to devote a long weekend of their precious time—with the expectation that they will learn something new and meet others with something useful to say—posed a challenge and a responsibility felt by all of us on the committee. According to the postconference survey, 93% rated the quality of the papers as well as the overall conference itself as "good" or "excellent." Our fellow committee members, the Board of Directors, and the conference staff should feel proud of the result.

One more word about the staff: For years TPRC has contracted with Economists, Inc., to provide administrative support for the conference. Since 1989 that support has been provided primarily by Dawn Higgins, assisted at critical times by Lori Rodriguez. While the committee gets to deal with the big picture, Dawn and Lori make sure that we know what needs to get done and by when, and then they make our decisions happen. We thank them again for making the conference appear to work seamlessly!

Ultimately, a conference derives excellence from its content. Thus most of the credit for the success of the 26th TPRC goes to everyone who submitted abstracts, followed up with papers, prepared presentations, and came to the conference ready to question, comment, and critique. TPRC has always been a great learning experience. As co-chairs of the committee, we are pleased to have had the opportunity to contribute to the advancement of telecommunications policy by leading the conference in 1998.

We are confident that you will find these papers highly readable, informative, and valuable.

Benjamin M. Compaine
Jorge Reina Schement

Introduction

Wishing someone "interesting times" can be a curse—but we feel blessed that telecommunications has experienced interesting times ever since we can remember. The past year brought exciting events and developments that are reflected in the chapters of this volume. Competition has progressed on various fronts both within the telecommunications infrastructure itself and in the services offered over it, highlighted by the growth in electronic commerce. At the same time, questions of market power have made front-page news, driven by a wave of mergers and the spectacular Microsoft antitrust case. Regulation is still with us and getting ever more complex as it tries to pave the way toward competition while also guarding new areas, such as privacy on the Internet. Regulation of universal service remains an evergreen topic. Convergence, talked about for more than a decade, finally appears to be arriving. Internet and telephone companies and telephone and cable TV companies have merged, and Internet telephony has become real as next generation telcos have deployed entirely packet-based infrastructures.

COMPETITION, MARKET POWER, AND ANTITRUST

The increase in competition in the telecommunications sector has been a worldwide phenomenon. It is expressed by the privatization of state-owned telephone companies, by deregulation of legal entry restrictions, and by the actual entry of newcomers in markets that have traditionally been dominated by a single firm. However, the road to competition is obstructed by old and new forms of market power, collusion, and government interference. In telecommunications, during the last few years, regulation has become an indispensable part of competition policy. This is strongly reflected in some of the following chapters, which describe regulatory policies ostensibly aimed at increasing competition.

The only chapter in this section on competition that is not directly related to regulation is Aron's and Wildman's "Effecting a Price Squeeze Through Bundled Pricing." It addresses an issue that is not limited to the telecommunications/information sector, although it is immensely topical here. *Bundling* is a contentious element of software competition that has been at the heart of the Microsoft antitrust litigation. Aron and Wildman develop a simple, yet powerful model of a monopolist in one product. This monopolist, through bundling, can profitably extend this monopoly to another product, for which it faces competition from a firm offering a superior product (in the sense that it would generate more surplus than the product offered by the monopolist). Bundling the two products turns out to be an equilibrium outcome that makes society in general and consumers in particular worse off than they would be with competition without bundling. Aron and Wildman show that simple antitrust remedies only work if the product for which the monopolist faces no competition is not essential for consumers to be able to enjoy the second product.

Aron and Wildman suggest that their model might fit the Microsoft case in the United States and other cases from the TV industry in Europe. Applied to the Microsoft case, their model would suggest that bundling is likely to be welfare reducing and that unbundling would not be a suitable remedy. Even if the model did not exactly apply to these cases it would provide illuminating insights for the economics of bundling, which is an area more known to be esoteric so far.

Spectrum auctions have the potential to increase competition in telecommunications markets because they help allocate scarce spectrum to those users most likely to create the largest benefits. Trivially, the very fact that additional spectrum is made available should increase competition, and spectrum auctions may facilitate freeing up such spectrum because governments value the funds thus generated substantially higher than government use (or nonuse) of that part of spectrum. Spectrum auctions, however, are quite complex and therefore may face problems that reduce their potentially beneficial effects on competition and the allocation of resources. Milgrom, in his chapter "Combination Bidding in Spectrum Auctions" addresses some of those complex issues in a way that is at the forefront and yet accessible to nontechnical readers. Specifically, Milgrom addresses the problems caused by simultaneous auctions of a particular frequency in many separate geographic areas. If valuations of these geographic areas were independent of each other, the valuation of a set of geographic areas would be simply the sum of the valuations of individual areas. However, interdependencies in valuations arise from, among other causes, complementarities in the use of equipment and marketing tools. As a result, winning a particular frequency band for a single area may be completely useless to a bidder who would value it highly in combination with other areas. To avoid ending up with unwanted areas, bidders would bid less aggressively. Milgrom, who has substantially influenced the auction design used in the recent Federal Communications Commission (FCC) auctions, analyzes com-

bined bidding designs that have good properties in getting the spectrum of combined areas to the bidders with high valuations and allow for straightforward
bidding. This is, however, a research subject with low chances of finding fully efficient auction designs. Milgrom therefore encourages researchers to conduct laboratory testing of practical proposals.

In his chapter on "Regulating Anticompetitive Behavior in the Internet Market:
An Applied Imputation Model for Peru," Briceño shows nicely the challenges
faced by regulators and antitrust authorities worldwide in the face of bottleneck
facilities provided by dominant operators. Telefónicá del Perú (TdP), as the dominant telephone network operator, only started to offer Internet services in 1997,
but in 1998 already had 36% of the market share for dial-up service and 57% of
the market share for dedicated Internet lines. Because this rapid expansion was the
result of aggressive pricing behavior, the suspicion arose that it was achieved
through low prices that TdP was charging itself for inputs that it was providing to
competitors at higher prices. Acting on the complaint of such a competitor, OSIP-
TEL—the Peruvian regulator—tried to establish whether TdP was charging itself
less for bottleneck inputs than it was charging competitors. Cost data provided by
TdP could not be directly verified, so OSIPTEL had to come up with independent
cost estimations using a bottom-up model of the Peruvian Internet service providers, based largely on publicly available data.

The results strongly suggest that TdP subsidized the supply of noncompetitive
inputs to itself. At the same time, TdP had somewhat inflated costs for competitive
inputs. (Alternatively, it could have used a different technology and made use of
economies of scope, for example, by collocating equipment.) OSIPTEL's findings
also suggest that TdP was charging competitors more than under the Efficient
Component Pricing Rule (ECPR). This would contradict the claim by some proponents of the ECPR that the ECPR would be the automatic outcome if pricing of
access were left to private negotiations rather than regulation.

Valuable trademarks have always conveyed market power to their owners and
erected entry barriers for others. This market power is often justified by superior
market performance and, in turn, trademarks induce superior performance because they only retain their value if consumer expectations are fulfilled. What
Mueller's chapter on "Trademarks and Domain Names: Property Rights and Institutional Evolution in Cyberspace" shows impressively is that, under current legal
interpretation, trademarks provide their owners substantial entry advantages on
the Internet. Identifying a trademark with a domain name simplifies advertising
and eases accessibility of the trademark owner's website.[1] Owners of trademarks

---

1. Today, users access websites by typing in domain names. In the future, there could be
mediators that would facilitate finding websites without the use of domain names. However, just like a store 50 yards off Main Street may have a poor location, mediators may not
appear to be good enough for trademark owners.

are advantaged in the registration of domain names in that other claimants are prevented from using letter sequences that resemble those used in the trademark. Mueller demonstrates empirically that extension of trademark protection to domain names violates competing and equally valid claims by others. He also shows that confusion arising from the use of letter sequences similar to those in the trademark is actually quite rare.

It is clear that the problem addressed by Mueller is going to stay important for some time, although most companies with trademarks already have established domain names. Given that, under realistic growth scenarios, the Internet will ultimately have to provide very large numbers of names, the similarity between names is bound to increase. A casual look into the Manhattan phone book will confirm that similar letter sequences based on proper names can be extremely common. Besides proper names, there are only a few thousand common words in the English language. Thus, domain names may have to use lengthy word combinations if similarities are to be avoided. But, because domain names have to be used in their exact letter sequence, it would not matter for communications purposes if they were approximately similar to each other. Thus, Mueller makes a convincing case that trademark rights should not be given priority in domain name registration.

REGULATION

What links Mueller's chapter with that by Srinagesh and Mitchell on "An Economic Analysis of Telephone Number Portability" is the value of ownership to an address in the virtual space of the telephone world. Local telephone number portability (LNP) creates a property right for users to their telephone numbers, rather than having the property right rest with the local exchange carrier (LEC). This property right assignment reduces the costs to subscribers of switching LECs and thereby has the effect of potentially increasing LEC competition. Srinagesh and Mitchell demonstrate other beneficial effects of LNP. Among these, the increased value of directories to incumbent local exchange carriers (ILECs) is one that could create a trade-off with the reduction in market power over local subscribers. Overall, the many social benefits of LNP have to be weighed against the substantial costs of its implementation. Such a cost-benefit analysis may have to wait until LNP has been fully implemented in at least one country. Even if it then turns out that net benefits have been negative, this result may not carry over to other countries, where costs of implementing LNP are likely to be smaller once the technology of doing so has matured. As in other areas of telecommunications, this is one where less advanced countries actually might experience advantages from being second.

The title of Brennan's chapter "Promoting Telephone Competition—A Simpler Way?" indicates that it seems to belong in the previous section on Competition, Market Power, and Antitrust. In fact, this again demonstrates the close link between regulation and competition in telecommunications. Brennan's paper is about deregulation, more precisely about the restriction on Bell operating companies (BOCs) to enter long-distance services in their regions. According to Section 271 of the Telecommunications Act of 1996, these restrictions can be lifted if certain complicated conditions on the presence of local competition in those regions are met. During the first 3 years of the Act, no BOC was able to overcome the hurdles set by 271, as implemented by the FCC. Brennan makes a seemingly simple suggestion to resolve the issues posed by BOC entry into long-distance services. He proposes to let the BOCs enter if and only if state regulators in the relevant geographic area have deregulated those BOCs. This suggestion is intriguing not only for its seeming simplicity but also for the issues it raises about the division of labor between federal and state regulators. As Brennan demonstrates, the suggestion is simple only on its surface, because a number of issues, such as those of interstate access pricing and the durability of state deregulation, have to be resolved with it. It also is simple only in the sense of delegation of authority. Once delegation has occurred, state regulators would have to design and implement rules for deregulation, something that may be as complex as Section 271.

Division of labor between regulatory levels, an issue raised by Brennan's chapter, is also taken up in the next chapter by Lehr and Kiessling. However, the two chapters come to quite different conclusions. Brennan argues in favor of decentralized decision making, whereas Lehr and Kiessling want to strengthen the central regulatory authority. The two views may be reconciled, though. What, in our view, Brennan wants to achieve with his proposal is two things: long-distance entry by the BOCs and their deregulation by state regulators. Thus, if Brennan's proposal were successfully implemented, state regulators would give up their regulatory authority over the BOCs. One of us editors had believed from the outset that forcing the state regulators to deregulate local telephone tariffs had been one of the prime, but unstated, goals of the Telecommunications Act of 1996. He thought that Congress had intended that but lacked the courage to put state price deregulation in the Act. He now believes that Congress simply did not anticipate all the forces that it let loose.

The chapter by Lehr and Kiessling, "Telecommunications Regulation in the United Sates and Europe: The Case for Centralized Authority," brings together an American and a European to provide fresh arguments for an old debate. Federal and state regulation have a long history in the United States, whereas the combination is new to Europe. In both regions, this is a time for intensive rethinking of the best approach. Both regions come from different constitutional positions, but the economic arguments presented by Lehr and Kiessling are similar. They postulate that the implementation of competition across different media is the main task of any regulatory authority in the telecommunications sector. Given this postulate, they argue that a centralized authority is going to be quicker and more effective in implementing competition. This argument seems to be borne out by the experience in the United States, where the FCC has been fighting for competition for the last 20 years, and in Europe, where over the last dozen years the European Commission has been instrumental in spreading telecommunications competition throughout the European Community. However, the United States also has a number of strong examples of progressive state commissions, such as New York and Illinois, that were showing the way toward local competition, and countries like the United Kingdom were leading the way in Europe. These forerunners, however, should not distract from the many laggards who need centralized authority to get going. Lehr's and Kiessling's case is further strengthened by the globalization of telecommunications issues through the Internet.

International impacts are also the perspective of Viktor Mayer-Schönberger. His chapter provides another European perspective on regulation. "Operator, Please Give Me Information: The European Union Directive on Data Protection in Telecommunications" lays out the European approach to data protection. In doing so, he not only shows how other countries, the United States in particular, may be directly affected by this Directive. He also challenges the approach taken by others. The Directive creates new regulations as a consequence of liberalization. The freer the markets for information and the easier the flow of information, the more important becomes the issue of data protection. Data protection is the flip side of data availability. One is only willing to provide data on oneself if the data can be protected. Thus, just as a brake helps a car to be fast, data protection may help to increase the flow of data.

Lawrence Lessig and Paul Resnick, in their chapter "The Architecture of Mandated Access Controls," deal with another, similarly important aspect of restrictions in the flow of information. They are concerned with the technology of restricting the flow of information over the Internet based on political, social, or legal mandates. Lessig and Resnick do not answer the question of whether such restrictions are morally or otherwise justified, but rather how such restrictions can be achieved effectively. They find their answer using a simple model of three actors: senders, intermediaries, and recipients. Because the current Internet architecture does not lend itself easily to content controls, the authors discuss regimes that would require modifications of the architecture, either using labeling or a database mapping Internet protocol (IP) addresses to jurisdictions. Such modifications will be costly and still provide only a highly imperfect basis for regulation. Thus, the question about the justifications for such regulations still needs to be answered. Can one leave the answer to the political process? Will governments do the right thing? The problem is that different jurisdictions have different valuations, some based on democratic principles, some not. Thus, there will be conflicts. The tools suggested by Lessig and Resnick may end up helping dictators rather than protecting minors from sexual exploitation. On the other hand, dictators will restrict access to information independently of such tools. When Germany was reunited in 1990, Western experts were surprised by the abysmal state of the former East Ger-

many's telecommunications infrastructure. It seems that by keeping that infrastructure underdeveloped, the regime could more effectively control speech than
they could by restricting speech in an advanced telecommunications system. Similarly, dictators will try to restrict the Internet with more repressive means if no
simple tools are available. Thus, although restrictive, the tools suggested by Lessig
and Resnick may actually help free the spread of Internet use.

Whereas regulation is usually constraining market participants, Global Information Infrastructure (GII) initiatives are actually trying to relieve constraints.
As such they are very appealing. However, as Hudson shows in her chapter on
"GII Project Initiatives: A Critical Assessment," they are marred by vague concepts and an emphasis on technologies rather than on their usefulness. Hudson
illustrates her claims through various partial initiatives that failed in developing
and developed countries. Are these just flaws of individual initiatives, or is the
GII concept flawed from the outset? Clearly, government initiatives compete
with private initiatives on a global scale. The Iridium project, for example, is a
private GII project. If it is successful, customers worldwide will benefit and
shareholders will get richer. If it fails, shareholders and creditors lose out. Thus,
the implementation of private GII projects is favored by two factors. First, the
risk is borne largely by the initiators. Second, the initiators overcome free-rider
problems by owning the project.

Evaluating experiences with government-led GII initiatives leads Hudson to
suggest improvements in the approach. She thus believes that the concept can be
salvaged. Even if they may not ultimately materialize, GII initiatives may have
functioned as a catalyst for national and international infrastructure projects that
are compatible with each other. The initiatives themselves may become fora for
cross-cultural understanding necessary in an area where global effects even derive
from domestic infrastructure projects.

UNIVERSAL SERVICE

Universal service is a persistent policy concern, yet at the same time an ill-de-
fined term. Convene a roomful of telecommunications policy researchers to discuss this topic and you will likely discover three very different sets of
expectations. The most politically minded folks will be ready to discuss justifications for universal and equitable access to telecommunications services and
causes and remedies for lack of achievement of that goal. Another group, identified by their affinity for the latest advances in telecommunications technology,
will want to know when a universal service policy will bring those advances into
their own homes. The final group, largely composed of economists, will expect
to discuss details of the cross-industry transfer payment systems created by federal and state regulators. Collectively, the three chapters in this section satisfy
each of these groups.

The first chapter in this section, "The Persistent Gap in Telecommunications: Toward Hypothesis and Answers" by Schement and Forbes, looks at universal service from a socioeconomic perspective. The authors observe gaps in aggregate telephone penetration rates among households of different races and ethnicities. They hypothesize that these gaps arise not just because of income differences but because of the compound effects of many variables including geography, housing type and ownership status, age and gender of head of household, and employment status. They find that available statistics are not broken down sufficiently to explain all of the gaps they find, some of which appear to be quite local in nature (e.g., large penetration differences between nearby counties within the same state regulatory system). Their discussion is at the same time a call for systematic data collection as the basis for multiple regressions that would identify the influence of socioeconomic factors on telecommunications penetration. This could help better target universal service policies and identify their success in increasing penetration levels of disadvantaged groups.

Schement and Forbes further ask whether such gaps are likely to persist as the services that are considered essential to full social, economic, and political participation in U.S. life migrate from the telephone to the computer. Drawing on historic penetration data for other communications media (such as radio and TV), the authors observe a difference between the essentially complete penetration of information goods (e.g., televisions) and the more variable penetration rates of information services (e.g., telephone bills that must be paid in full each month, do not generate a secondhand market, etc.). They discuss evidence from Hispanic households suggesting that personal computers fall more into the goods category, so that any gaps in penetration of PCs will eventually close of their own accord.

Internet access, on the other hand, falls more into the service category. The second chapter in this section, "Do Commercial ISPs Provide Universal Access?" by Downes and Greenstein, looks not at penetration rates among users, but at the availability of competitive Internet access service in different geographic areas. The authors make innovative use of lists—publicly available on the Internet—of dial-up Internet points of presence, which they correlate with geographic information from the U.S. Census. They find that even without an explicit universal service policy for Internet access, over 99% of the U.S. population has access to at least one Internet service provider (ISP) via a local (toll-free) call, and over 96% of the population has access to three or more. The areas with the least service are overwhelmingly rural, suggesting at least some economies of scale that are not satisfied by the low densities of very rural areas.

The results of this study provide a useful input to the policy debate. However, it would be a mistake to conclude that the commercial ISP market has achieved near universal service completely of its own accord. As the authors caution, parts of the Internet's infrastructure were effectively subsidized because they were put in place by the U.S. government prior to commercialization of the Internet in 1995. Even more basic, universal availability of dial-up Internet access could not exist without universal availability of telephone service of sufficient quality to support the use of modems. To the extent that universal service policies have influenced the spread of digital technologies in the phone network, they have indirectly made universal availability of Internet access a more easily achieved goal.

"Proxy Models and the Funding of Universal Service," by Gabel and Kennedy, deals with a specific issue in the economists' category characterized earlier. In the Telecommunications Act of 1996, the U.S. Congress substantially extended the scope of universal service policy and tried to set up explicit financing of universal service burdens, moving away from implicit cross-subsidies contained in local access charges paid by long-distance carriers. Universal service still involves a system for transfer payments from telephone service providers and users in lower cost regions of the United States to providers and users in higher cost areas. This cost averaging is intended to perpetuate the lack of significant geographic differences in pricing for local telephone service, a uniformity that has been a feature of the U.S. telephone system since its regulated monopoly days. The underlying rationale is to avoid the risk that subscribers in high-cost—predominantly rural—areas will be forced off the telephone network because of high prices, and to reduce burdens from price restructuring that local competition might place on some customers.

As part of its implementation of the Act, the FCC is trying to determine the actual costs of any particular "high-cost" operator. Because this determination clearly influences the magnitude of the transfer payment received by the operator, the regulator has a pressing need to validate cost figures claimed by any operator. At the same time, such validation is prohibitively expensive to do for each high-cost area. The FCC is therefore considering so-called cost proxy models that combine industrywide cost data with local geographic information.

Gabel and Kennedy deal with a weak point of such cost proxy models. Those models can, in principle, be quite accurate when it comes to standardized network components, such as digital switches. However, unfortunately, as the chapter explains, when it comes to outside plant, such as cable installation or poles, costs are not standardized and, generally, not publicly available. The various cost models used by the FCC are therefore not as applicable as they might be, because of their reliance on expert opinion that is itself difficult for regulators to validate. Gabel and Kennedy offer an alternative approach. Relying only on public domain data about telephone network construction costs, they have created, and made available on a website, a linear regression model that allows any interested party to plug their local conditions into the model and extract a likely estimate of cost.

Despite the authors' many qualifications and caveats, their method appears reasonable for universal service determinations. Of more concern is its lesser applicability to unbundled network element (UNE) proceedings, which apply to all areas of the country, not just the rural areas for which the authors were able to obtain public data. The lack of publicly available cost data remains a persistent concern in tele-

xxii

communications policy research. However, the authors' innovative way to construct public data may perhaps spur others to think creatively about the UNE data problem.

CONVERGENCE

We use the term *convergence* to refer specifically to the coming together of public voice and data networks. Initially, that meant data transport over voice channels. Now, it increasingly means that voice, rather than being a service inextricably bundled with a particular network (the public switched telephone network [PSTN]), becomes one of many services provided over data networks such as the Internet. In this convergence, the optimality of connection-oriented (circuit-switched) technology for voice service gives way to the flexibility of connectionless (packet-switched) technology that enables a wide variety of services. The range of services includes traditional Internet applications such as e-mail and the Web; services like telephony, radio, and television that have traditionally been provided over customized networks but are now appearing, in evolving forms, on the Internet as well; and new services not yet dreamed of but sure to emerge eventually.

The four chapters in this section remind us that we should not expect this transition to proceed at the same pace across all aspects of the telecommunications system. Internet telephony—defined as the carriage of telephone calls over IP-based networks—poses different sets of challenges and rewards for long-distance network operators (the first chapter), local access network operators (the next two chapters), and, perhaps most vexing of all, regulators (the final chapter).

Commercial announcements to date of Internet telephony network deployments, such as those from Qwest and Level 3, have mainly focused on the long-distance market. The chapter by Weiss and Hwang, "Internet Telephony or Circuit Switched Telephony: Which Is Cheaper?" sheds some light on why this may be so. Based on detailed models that the authors construct of switching and interoffice transmission facilities in both circuit-switched and Internet telephony architectures, they conclude that the costs of these components are significantly lower in the Internet telephony architecture. In their model, the cost reduction stems not—as is often supposed—from differences in regulatory treatment or from the statistical multiplexing of bandwidth that packet switching allows, but rather from the greater compression of voice signals made possible by Internet telephony, in comparison to the fixed 64 Kbps channels of the traditional phone system. This result is consistent with the consensus of the conference's plenary panel on "Netheads vs. Bellheads," as well as the research paper presented by Andrew Odlyzko of Bellcore,[1] that the Internet architecture wins more on the basis of flexibility than cost cutting.

1. See "The Internet and Other Networks: Utilization Rates and Their Implications" at http://www.research.att.com/~amo/doc/internet.rates.pdf

Assuming Weiss and Hwang are right, we can expect Internet telephony to follow the pattern of other telecommunications innovations, such as digital transmission: Long-distance comes first, whereas local access, because of the high cost of the installed base, proves much more challenging. The next two chapters each address different aspects of this local access problem. The first, "An IP-Based Local Access Network: Economic and Public Policy Analysis" by Fryxell, Sirbu, and Wanichkorn, contributes a proof-of-concept design for an IP-based central office and compares its projected costs against the Hatfield model for PSTN access. In contrast to today's central offices, which connect subscribers' analog local loops to digital but still voice-oriented circuit switches, the IP central office connects digital local loops (provided using digital subscriber line [DSL] technology over existing copper wiring), through DSL multiplexers and asynchronous transport mode (ATM) switches, to IP routers. Today's architecture is oriented toward voice as the dominant application, with data an "add-on" service provided through modems. In the IP-based central office, data provision is the primary service, with plain old telephone service as the add-on. This transition is hidden from analog phone users by special IP gateways that provide back compatibility by connecting analog local loops to IP routers. The circuit switch is no more.

Modeling the cost of this design proves challenging, as several of the necessary components are not yet available commercially, and several assumptions must be made for the sake of comparison. However, the authors do not find these assumptions critical to their results. Rather, they find a key sensitivity to the number of phone lines needed by the customer. In the IP-based architecture, multiple phone lines can be provisioned over the same physical circuit using DSL technology. The trade-off, however, is higher costs of customer premises equipment (CPE) and multiplexing equipment at the central office. As a result, the IP-based architecture is only cost-effective today for customers needing multiple lines, more typical of small businesses than most residential users. However, as both the CPE and multiplexing equipment are based on digital technologies experiencing the price-performance improvements typical of the computer industry, we should expect this break-even point to shift downward over time.

As the authors note, however, the services provided by the two architectures are not equivalent. The PSTN architecture benefits from higher reliability; powering that is independent of the local electricity supply; and established procedures for operational support, number space management, and directory services. On the other hand, the IP architecture provides "always on" and higher speed access to data services. For some customers, these benefits would outweigh IP's drawbacks and justify its higher cost. For just how many customers and at how much higher cost remain open questions that the marketplace could test—if given the chance.

Whether the local access market is likely to be competitive—assuming the market is for always on, high-speed Internet access, not just plain old telephone service—is the subject of Clark's chapter, "Implications of Local Loop Technology for Future Industry Structure." Clark performs a thought experiment in which he envisions how the intertwined evolution of new Internet-based services, new local loop technologies, and new consumer equipment may affect the level of competition experienced by customers. He concludes that in areas currently served by both telephone and cable operators (which includes most of North America and several other industrialized countries), duopoly is the likely outcome, as consumers express their preference for higher speed, always on access, and dial-up access providers are forced into an ever smaller share of the market. This outcome would represent a significant reduction from today's broad consumer choice among ISPs.

Aside from regulatory intervention, Clark raises the option that research targeted at higher bandwidth wireless access might mitigate the duopoly outcome. The local access problem is, as its name implies, a very location-specific one. The more technical options exist, the more likely any given community is to find a solution that works for local circumstances. In contrast, regulatory approaches are more likely to impose homogeneous solutions on heterogeneous situations.

Finally, Weinberg's chapter, "The Internet and 'Telecommunications Services,' Access Charges, Universal Service Mechanisms, and Other Flotsam of the Regulatory System," delves into the regulatory question in detail. He examines the effects of convergence on regulation and vice-versa. Weinberg analyzes the politically charged question of whether Internet telephony service providers should be included in the systems of federal Universal Service Fund subsidies and interstate access charges that apply to public operators of telephone services. By the language of the U.S. Telecommunications Act of 1996, such decisions hinge on whether the FCC classifies Internet telephony as a telecommunications or information service. Weinberg shows how spurious these distinctions are when applied to Internet telephony. In a series of humorous clarifying examples, he shows how, with the ease of a software upgrade, a service's classification could instantly change from one category to the other.

Weinberg argues for a regulatory approach that more closely mirrors the technology layers that define the Internet's architecture. He proposes imposing universal service obligations only on the owners of physical communications facilities. Although such an approach would indeed simplify the FCC's decision process, it raises more questions than one chapter can answer. Would this approach require new legislation or new interpretations by the FCC? What other effects might such a redefinition have—intended or not (e.g., might it have the paradoxical effect of reducing the incentives to own facilities, therefore discouraging the construction of competing networks)? Weinberg does discuss the issues surrounding how such an obligation might be implemented. Most important, he raises a question that is all too often missing from discussions of existing regulation: What legitimate policy goal do these regulations serve, and does it still make sense in the new environment?

Ingo Vogelsang
Sharon Eisner Gillett

COMPETITION

Effecting a Price Squeeze Through Bundled Pricing

Debra J. Aron
LECG, Inc.

Steven S. Wildman
Northwestern University and LECG, Inc.

We show that a monopolist over one product can successfully leverage that monopoly power into another market so as to exclude rivals from that market by using bundled pricing. This strategy is part of an equilibrium in which the monopolist behaves rationally, both in the short run and the long run. Indeed, the exclusionary pricing is profitable in the short run, but would not be profitable in the absence of the competitor. We show that the bundled pricing unambiguously generates welfare losses, both in terms of social welfare and in terms of consumer surplus. Our model is applicable to a variety of industries, including pay television and computer software. The analysis is particularly relevant to the current debate over Microsoft's bundling of its browser with its operating system. We examine the potential for requiring the sale of unbundled elements (without also precluding bundling) as an antitrust remedy and find that such a requirement, coupled with an "adding up" rule, is an effective remedy in some cases. However, additional remedies are likely to be needed in the Microsoft case to address the special demand characteristics of new buyers of computers.

INTRODUCTION

One of the most significant decisions facing antitrust authorities in the 1990s will be how they choose to resolve their investigations of Microsoft. One critical element of the behavior at issue is Microsoft's bundling of its browser with its operating system. Microsoft's new operating system, Windows98, has an Internet browser tightly integrated into it. Unless the Department of Justice forces Microsoft to provide the products separately (which Microsoft has claimed it cannot do), customers will not be able to buy the operating system without the browser, nor will they be able to buy the browser without the operating system. It is generally accepted that Microsoft holds an effective monopoly in the market for operating systems for personal computers. Competitors claim that by bundling the browser with its monopoly operating system, Microsoft makes it impossible for other browser firms to compete in the browser market. This not only has an exclusionary effect on competition but, it is claimed, a chilling effect on innovation.

Opinions regarding the anticompetitive nature of Microsoft's behavior and what might be done about it vary from those who believe that Microsoft is simply an aggressive, successful, innovative competitor that should not be punished for being successful, to those who believe that nothing short of a breakup of Microsoft will remedy the anticompetitive effects of its actions. The positions have assumed something like a religious fervor, in part because of the enormous amounts of money at stake, and in large part because there is remarkably little economic theory informing the debate.

In fact, the economic issue in the Microsoft case arises in other industries as well. In the pay television industry, it is sometimes claimed that a provider who owns monopoly rights to an anchor, or "marquee," channel can preclude competition in thematic channels (such as comedy or science fiction channels) by bundling their own thematic channels with the anchor. The idea is that consumers value some channels (such as HBO or CNN) so highly that those channels drive consumers' decisions regarding which program platform they will buy. Conversely, the logic goes, a provider that attempts to compete by offering a thematic channel on a stand-alone basis, without an anchor channel, would not be able to survive the competitive pressure of a rival with an anchor.

This argument, that anchor programming is critical to the viability of a program provider, is not only intuitively appealing, but appears to be borne out by the recent history of pay television in Britain. In that industry, attempts to compete with the dominant provider of pay television, BSkyB, have been failing miserably.[1] One conjecture explaining the failure of competitors is the fact that BSkyB controls most of the critical programming rights in Britain, enabling it to use bundled pricing to execute a price squeeze against rivals. Hence, like the Microsoft case, the claim in the pay television industry is that a firm that monopolizes one product (here, an anchor channel) can effectively leverage that monopoly to preclude competition in another product market by using bundled pricing.

An (unsuccessful) antitrust claim by Ortho Diagnostic Systems, Inc. against Abbott Laboratories, Inc. under Section 2 of the Sherman Act (among other

---

1. "A Massive Investment in British Cable TV Sours for U.S. Firms," *The Wall Street Journal,* December 17, 1997.

claims)[2] provides a third example. Abbott and Ortho both produce blood screening tests that are used to test donated blood for the presence of viruses. Abbott produced all five of the commonly used tests, while Ortho produced only three of the five. The claim centered on the prices in contracts that Abbott negotiated with blood centers. Ortho claimed that those prices bundled the five tests together in such a way that Ortho was precluded from competing in the tests that Ortho could provide in competition with Abbott. Because Abbott was effectively a monopolist in two of the tests, Ortho claimed that Abbott could and did use a bundled pricing strategy to leverage its monopoly into the other nonmonopolized tests and preclude competition there.

While it may seem obvious that a monopolist can preclude competition using a bundled pricing strategy, what makes the analysis less obvious is the need to demonstrate that by doing so, a monopolist does not incur costs in excess of the profits it might realize in the newly monopolized market. For a bundled price squeeze to be a significant antitrust concern, it is not sufficient that it be possible for a monopolist to execute the squeeze; it must also be rational for the monopolist to do so, and it must be part of an equilibrium in which the competitor also behaves rationally. Moreover, to be an antitrust concern, doing so must entail a decline in social welfare or, at least, in consumer surplus.

In this chapter we demonstrate that, under plausible conditions, it is indeed possible in equilibrium for a provider who monopolizes one product (or set of products) to profitably execute a fatal price squeeze against a rival in another product by using a bundled pricing strategy. We show that if all consumers demand a single unit of the monopolized product and have identical preferences, bundling cannot be sustained as an equilibrium leveraging strategy. However, a downward-sloping demand for the monopolized product may be sufficient to support bundling as a strategy for leveraging monopoly from one market to another. The ability to drive or preclude a rival from a market using bundled pricing exists even if the rival can offer a superior alternative to the monopolist's nonmonopolized product. We demonstrate this in the context of a game theoretic model of competition between providers in which both providers and consumers are assumed to act rationally. That is, the monopolist in one market can squeeze the rival out of the second market not by making unsustainable threats or by offering below-cost pricing, but by offering a profitable bundled pricing scheme against which the rival cannot compete.

The three examples just offered have the common feature that the monopolist is thought to have the ability to leverage its monopoly in one or more products into a potentially competitive market. The Microsoft case differs from the other two, however, in an important respect. In the pay television industry, when customers

---

2. Ortho Diagnostics Systems, Inc. v. Abbott Laboratories, Inc., 920 F. Supp. 455 (S.D.N.Y. March 15, 1996) and Ortho Diagnostics Systems, Inc. v. Abbott Laboratories, Inc., 926 F. Supp. 371 (S.D.N.Y. April 26, 1996).

subscribe to a thematic channel, they could enjoy it even if they did not subscribe to the anchor channel. Similarly, a blood screening center could make use of the test for hepatitis even if it did not have the test for HIV. In contrast, customers cannot operate a browser without having an operating system on their computer. Hence, the strategies that are available to alternative browser providers are more limited than those available to providers of thematic channels or competitive blood tests. Some customers already have an operating system and are in the market to buy a browser; and some customers are buying a new computer and must therefore buy an operating system. For the first group, which we consider the "standard" case, the customers' decision problem is similar to the cable TV subscriber's decision regarding channel subscription. For the second group, however, which we call the "new buyer" scenario, the decision differs because the customer cannot choose to buy only a browser and no operating system. Hence, in the latter case, the browser competitor cannot hope to induce customers to abandon the monopolized product entirely. Strategically, this strengthens the monopolist's ability to leverage. In the model that follows, we fully examine the standard case, in which consumers would be willing to buy the competitor's browser (or cable television channel, or blood screening test) even in the absence of the monopolist's product. We show that even in that case, bundled pricing can be used to exclude competitors. We also briefly address the new buyer scenario. It is straightforward to show that the existence of the competitive alternative is not a constraint on the monopolist's monopoly power, as it is in the standard case. This limits the feasibility of simple pricing remedies.

Our chapter is organized as follows. The next section develops an example that illustrates the intuitive logic behind our results. The assumptions of our formal model are presented after that, and our main result demonstrating the anticompetitive effects of bundling is presented in The Pricing Game. Nontechnical readers can skip the Assumptions and The Pricing Game sections because the intuition is contained in The Logic of Anticompetitive Bundling: An Example.

We then contrast the bundling equilibrium with the market outcome when bundling is precluded by regulation, and next compare the social and consumer welfare under the bundled and stand-alone pricing outcomes. We show that welfare is higher when bundling is precluded, both in terms of consumer surplus and total social welfare.

In the subsequent section, we consider whether stand-alone pricing together with bundled pricing ("mixed bundling") is a viable remedy, and conclude that, in the standard bundling scenario, mixed bundling is a remedy if it is accompanied by the requirement that the stand-alone prices sum to no more than the price of the bundle (the "adding-up" rule). In the section after that, however, we examine the new buyer scenario, in which customers who do not buy the monopoly product (e.g., the operating system) have no use for the secondary product (the browser). In this case, the anticompetitive effects of bundling are even stronger than in the standard case, and mixed bundling is not a viable remedy. The last section contains concluding comments, including the fairly straightforward implication of our results that bundling competitive products with monopoly products can discourage innovation by firms other than the monopolist.

THE LOGIC OF ANTICOMPETITIVE BUNDLING: AN EXAMPLE

Our model assumes that there are two potential firms in the industry. Firm 1 owns an operating system (or anchor channel or any monopoly product), A, and also provides a second product, such as an Internet browser or a thematic television channel, B. The other firm, Firm 2, provides a product that is a substitute for B, which we will call B^*. Firm 2, however, has no substitute for channel A. Hence, Firm 1 can provide A and B in a bundle, whereas Firm 2 can only providé B^* on a stand-alone basis.

Although it might appear obvious that Firm 1 can force Firm 2 out of the market by bundling A and B, on closer examination the ability to do so is more delicate. If the price of the bundle is too high, consumers would prefer to buy only B^* rather than the A and B bundle, even though they value A very highly. That is, at a high enough price for the bundle, consumers will simply get more from buying B^* at a reasonable price than from purchasing A and B at the high price. Taking this into account, Firm 1 cannot set a bundled price that Firm 2 can simply undercut and thereby attract all the customers. Such a bundled price would benefit, rather than squeeze, Firm 2.

On the other hand, if Firm 1 sets a very low price to drive Firm 2 out of the market, Firm 1 forgoes profits. If it were more profitable to Firm 1 to set a higher price at which both firms could coexist, then the squeeze strategy would not be rational, would not be a credible threat, and therefore would not be an equilibrium in the market. By showing that the price squeeze is, in fact, an equilibrium, we are showing that no other pricing strategy would generate higher profits for Firm 1, given rational behavior by Firm 2, and that Firm 2 cannot avoid the squeeze.

The model assumes that there are two consumers, X and Y, who differ in their valuations of the monopoly product.[3] The difference in valuations is critical because, surprisingly, it is what makes bundled pricing powerful. The conditions that make bundled leveraging profitable are weaker, however, than those required for a two-product monopolist to find it profitable to bundle the products together. To illustrate the point before we introduce the full model, consider the following example. To make it more concrete, suppose that the relevant products are two types of pay television channels.[4] Overall, Customer X likes watching television more

---

3. The model is easily generalized to many consumers of each type.
4. See Owen and Wildman (1992, chap. 4) for an analysis of channel bundling by cable television system operators.

than does Customer Y. Let A be an anchor movie channel, and B be a science fiction channel, both supplied by Firm 1. Product $B*$ is a second science fiction channel supplied by Firm 2 in competition with B. Although it competes in science fiction channels, Firm 1 is a movie channel monopolist and faces no direct competition to A. Both X and Y prefer watching the movie channel to watching the science fiction channel. However, X likes the movie channel better than does Y; conversely, Y enjoys watching the science fiction channel more than does X.

To quantify the example, let Customer X value A at \$10 and B at \$3; let Customer Y value A at \$8 and B at \$4. The total value ascribed to A and B by X is \$13; the total value for Y is \$12. Forget about Firm 2 and $B*$ for the moment, and suppose Firm 1 is a monopolist in A and B. Just considering these valuations, it is straightforward to see that if Firm 1 had to price each product separately, it could do no better than to set the price of A at \$8 (at which both X and Y would buy A), and the price of B at \$3 (at which both X and Y would buy B). Revenues from each customer would be \$11. However, if the firm could set a bundled price for A and B together, and did not offer A and B separately, it could do even better by pricing the bundle at \$12. At \$12, both X and Y would buy the bundle, and revenue per customer would be \$12 rather than \$11. It is not possible to achieve a revenue per customer of \$12 with stand-alone pricing.

What makes bundled pricing profitable for the two-product monopolist in this example is the negative correlation in the ordering over which customer has a higher valuation for each product. In our example, X is the higher valuation customer of the movie channel but the lower valuation customer of the science fiction channel. Y is the higher valuation customer of the science fiction channel but the lower valuation customer of the movie channel. When customers face a bundled price, it is only their total valuation of the two products together that matters, rather than the individual components of the valuation. As long as stand-alone pricing is also an option, not having to satisfy consumers' individual valuations of the channels gives the provider more freedom in pricing.

The observation that bundling may be profitable when demands are negatively correlated is not new. It was first suggested by Stigler (1963) as an explanation for block booking by movie distributors. Adams and Yellen's (1976) paper was probably the most influential statement of this principle. Subsequent work has shown that it is only necessary that different consumers' demands for two products be independently distributed for bundling to be more profitable than stand-alone pricing in a variety of circumstances (see, e.g., Bakos & Brynjolfsson, 1998; McAfee, McMillan, & Whinston, 1989; Salinger, 1995). In fact, bundling may even improve profits when there is some degree of positive correlation of demands among consumers—a possibility illustrated with an example by Adams and Yellen.

In his analysis of tying as an exclusionary device, Whinston (1990) showed that the same logic that might lead a monopolist of two goods to sell them as a bundle instead of independently at their stand-alone profit maximizing prices might also lead a firm with a monopoly over a single product to bundle that product with a second in which it faces at least the prospect of competition. In Whinston's model, the incentive to bundle comes from variation in the intensity of consumer preferences, as measured by reservation prices,[5] for the monopolized product that is independent of preference variation in the residual demand faced by the monopolist for the competitive product.[6] For this situation, Whinston demonstrates the possibility of "innocent" exclusion, in which a monopolist chooses to offer both products as a bundle simply because bundled pricing is more profitable than standalone prices given the residual demand it faces for the competitive product. Because bundling denies competitors the possibility of sales to some of the customers for the monopoly product, the competitors may be reduced to serving a portion of the market too small to sustain them—hence the possibility of exclusion. The fact that the exit of competitors is not a factor in the monopolist's decision to bundle in this situation is what makes it "innocent."

In this chapter we show that bundling may be a profitable exclusionary strategy under demand conditions considerably less restrictive than those assumed in the traditional two-product monopoly models and their application to tying by Whinston. In particular, we show that it is only necessary to have variation in preference intensity (as measured by reservation demands) for the monopoly product alone to make bundling it with a competitive product a potentially profitable strategy for leveraging monopoly from one market to another. Further, although the strategy involves no sacrifice of short-term profits, as in models of predation, it also need not be "innocent" in Whinston's sense for it to be profitable. We show that a monopolist may bundle in equilibrium only because of the competitive threat; that is, it would bundle in cases where it would not bundle if faced with the same demand in the absence of the competitor.

These claims are easily illustrated with a slight modification of the preceding example. Let X value A at \$10 and Y value A at \$8, as before; but now assume X and Y both value B at \$4. For a monopolist over A and B, bundling A with B now adds nothing to revenue beyond the total of \$24 that would be realized with standalone monopoly prices of \$8 for A and \$4 for B, since the maximum price at which both X and Y buy the bundle would be \$12. Suppose now that a new firm (Firm 2) enters offering a superior version of B, $B*$. Customers X and Y both value $B*$ at \$5. Further assume that both firms incur a fixed cost of \$3 to produce their respective versions of B, but per-unit costs are zero. Thus, because it produces a better version of B for the same cost, the entrant is more efficient than the incumbent.

---

5. A *reservation price* is the highest price a consumer is willing to pay for a product without foregoing consumption of the product entirely.

6. A firm's *residual demand* is the demand for its product that remains after its competitors' sales have been taken into account.

Similarly, suppose the incumbent, Firm 1, incurs a fixed cost of $3 to produce A and no per-unit costs.

It is easy to see that there is no price for $B*$ high enough to cover Firm 2's costs that cannot profitably be undercut by the incumbent selling A and B as a bundle. Suppose the price of $B*$ were set at $3. Buyers of $B*$ would realize consumer surplus of $2. At a price of $9.99 for the bundle, X would enjoy consumer surplus of $4.01, and Y's surplus would be $2.01. Hence, both would buy the bundle. The entrant would have to lower the price of $B*$ to $0.98 to win X from the incumbent against the price of $9.99 for the bundle. The entrant would also get customer Y at this price; but $2 \times \$0.98$ is less than the $3 needed to cover the cost of $B*$. Moreover, no other price below $3 will enable the entrant to cover cost either.[7] There is also no price above $3 that will permit the entrant to make profitable sales, because if the incumbent can profitably undercut the $3 price for $B*$, any higher price can also profitably be undercut. Thus, bundling A with B enables the incumbent profitably to exclude a more efficient rival.

At the bundled price of $9.99, the incumbent earns profits of $(2 \times \$9.99) - (2 \times \$3) = \$13.98$. The entrant sells nothing and earns profits of zero. Now, if the incumbent were precluded from offering bundled prices, the best it could do is sell only A at $7.99, whereas Firm 2 could sell $B*$ at $2.49 to both customers.[8] The incumbent would make profits of only $12.98, a loss of $1 compared to the bundled strategy, and the more efficient firm would win the competition to supply the secondary product $B*$.

To see why bundling is not an "innocent" strategy (in Whinston's terms) in this situation, suppose instead that Firm 1 is a monopolist in both products and that the maximum that Customers X and Y are willing to pay for B is $1.99. Clearly the monopolist would be indifferent between offering A and B separately at prices of $8 and $1.99, respectively, or selling the two as a bundle for $9.99. Revenues and costs would be the same in both situations. Note, however, that from the perspective of the monopolist, the demand for B in this situation is identical to that described before when consumers valued B at $4, but $B*$ was available for $3, which they valued at $5. The residual demand faced by Firm 1 would be characterized by a reservation price for B of $1.99 for X and Y. Thus, the greater profitability of bundling over stand-alone pricing in the former situation derives not from any negative correlation between demand for A and residual demand for B, but entirely from the fact that bundling prevents Firm 2 from being able to compete effectively for sales to Customer X. Moreover, as we pointed out earlier,

---

7. At or below $0.98 both customers will buy. Between $0.98 and $2.99, only Y will buy. Above $2.99, no one will buy. Hence, no price below $3 would generate enough revenue to cover costs.

8. The monopolist would have to respond with a price of $1.48 to make any sales of B, but then it would not be able to cover its cost of $3.

Firm 1 would gain nothing from bundling, given these valuations for A and B, in the absence of Firm 2.

Our assumptions on the cost structure of the industry in our more general model are, in our judgment, quite benign. We assume that the marginal cost of providing service to customers is constant and, without loss of generality, we set the constant equal to zero. We also assume that there is a cost C incurred by any firm that actually provides a product to customers. Cost C is sunk and fixed, but C need not be incurred if the firm learns that it will have no customers for its product, and therefore will not provide it. In subscription television, C may reflect the costs of securing satellite carriage; in software, C may reflect the costs of setting up distribution channels or the costs of debugging the program.

The formal model in the next three sections shows that, subject to certain restrictions on the difference in consumer valuations of the monopoly product, the degree of superiority of $B*$ over B, and the size of C, there is always an equilibrium in which the provider of the monopoly product can use bundled pricing to squeeze the rival out of the market. The ability to exclude a rival holds even if the rival has a more attractive alternative to B. Moreover, we show that the bundled price squeeze decreases social welfare and net consumer surplus, relative to the equilibrium without bundled pricing.

ASSUMPTIONS

The model is constructed as follows:

A. Define two firms, Firm 1 and Firm 2.

B. Define two consumers, X and Y.

C. Firm 1 sells product A and product B.

D. Firm 2 sells product $B*$, which is a substitute for B. In particular

$$U_i(A, B, B*) = U_i(A) + \max\{U_i(B), U_i(B*)\}, i = X, Y,$$

where is customer i's utility function. For simplicity of notation, let $U_i(j) = i_j$, for $i = X, Y$ and $j = A, B, B*$.

E. Without l.o.g., let $X_A \geq Y_A$.

F. Let $\Delta \equiv X_{B*} - X_B = Y_{B*} - Y_B$.

G. Each firm incurs a fixed cost of C per product (not per unit of the product), but C is incurred only after the firm observes how many customers it has; if it has no customers it need not incur C.

Firms make simultaneous price offers. For Firm 1, the price may be a price of the bundle of A and B, denoted $\hat{P}$, or separate prices for A and B, denoted P_A and P_B, respectively, or all three. For Firm 2, the price is the price of B^*, denoted P_{B^*}. After price offers are made, each consumer purchases from the firm giving him or her the highest net consumer surplus. We assume that if customers are indifferent between two options, they choose each with equal probability. Each firm incurs C if it has one or more customers with positive probability. We look for a Nash equilibrium to this game.

THE PRICING GAME

We assume prices can vary only in discrete units, with $\varepsilon > 0$ being the smallest unit, ε small relative to C.

Proposition 1: Assume the following conditions:
  1. $2(X_A - Y_A) > C$

  2. $2Y_A - X_A - \varepsilon > 0$

  3. $X_A - Y_A < \varepsilon$

Then the following are Nash equilibria in bundled pricing:

$$\left\{ \hat{P} = Y_A - \Delta + P_{B^*} - \varepsilon, P_{B^*} \in \left(\Delta + \tfrac{c}{2}, C + \varepsilon \right] \right\} \text{ for } \Delta \leq \tfrac{c}{2}.$$

Proof: In this equilibrium, both X and Y will purchase the bundle from Firm 1: If X purchases the bundle, X gets net consumer surplus (NCS) of

$$X_A + X_B - (Y_A - \Delta + P_{B^*} - \varepsilon) = X_A - Y_A + X_{B^*} - P_{B^*} + \varepsilon.$$

If X purchases B^*, X gets NCS of $X_{B^*} - P_{B^*}$. Hence, X will purchase the bundle, by Assumption E.

If Y purchases the bundle, Y gets NCS of

$$Y_A + Y_B - (Y_A - \Delta + P_{B^*} - \varepsilon) = Y_{B^*} - P_{B^*} + \varepsilon.$$

If Y purchases B^*, Y gets NCS of $Y_{B^*} - P_{B^*}$. Hence, Y will purchase the bundle. In this equilibrium, Firm 1's profits are

$$\pi^1 = 2(Y_A - \Delta + P_{B^*} - \varepsilon) - 2C > 0 \text{ by } P_{B^*} > \Delta + \tfrac{c}{2}.$$

Firm 2's profits are $\pi^2 = 0$.

To verify that this set of prices is an equilibrium, we first verify that if Firm 2 sets

$$P_{B*} \in \left(\Delta + \tfrac{C}{2}, C + \varepsilon \right],$$

Firm 1 can do no better than $\hat{P} = Y_A - \Delta + P_{B*} - \varepsilon$. If Firm 1 sets $P = \hat{P} + \delta + \varepsilon$, for $\delta > 0$, customer Y would defect to B^*. If $\delta > X_A - Y_A$, X will also defect, and clearly this cannot be a profitable deviation. Hence, let $\delta > X_A - Y_A$. Firm 1's profits would be $\hat{P} + \delta - 2C$. This would exceed π^1 only if $\delta > Y_A - \Delta + P_{B*} - \varepsilon$. But this is not possible by Condition 2. Hence, Firm 1 cannot benefit from setting $P > \hat{P} + \varepsilon$.

If Firm 1 set $P > \hat{P} + \varepsilon$, X would purchase the bundle and Y would be indifferent between the bundle and B^*. Assuming Y purchases the bundle with probability θ, Firm 1's profits would be $(1 + \theta)(\hat{P} + \varepsilon)$. This can exceed π^1 only for θ close to 1. We assume that $\theta = .5$; hence $P = \hat{P} + \varepsilon$ is not a profitable deviation. (If $\theta \approx 1$, the equilibrium bundled price would be simply $P = \hat{P} + \varepsilon$, instead of $\hat{P}$).

Clearly, Firm 1 cannot profit by deviating to a $P < \hat{P}$, because in the candidate equilibrium, Firm 1 gets both customers at $\hat{P}$.

We must also rule out deviations to nonbundled pricing. That is, do there exist prices P_A, P_B, that yield Firm 1 a higher profit than π^1 against P_{B*}. First, consider candidate prices P_A, P_B, such that both customers purchase A and B from Firm 1. P_B must be less than $P_{B*} - \Delta$ or X and Y will purchase B^* at P_{B*} (or purchase B with probability .5). Also, P_A must be less than Y_A or Y will not purchase A with probability 1 (if Y purchases A, X will too, because $X_A > Y_A$). Hence, the relevant constraints are $P_A < Y_A$ and $P_B < P_{B*} - \Delta - \varepsilon$. The highest prices that satisfy these constraints are $P_A = Y_A - \varepsilon$, and $P_B = P_{B*} - \Delta - \varepsilon$. Firm 1's profit would be $2(Y_A + P_{B*} - \Delta - 2\varepsilon) - 2C < \pi^1$. Hence, this is not a profitable deviation.

Next, consider deviations in which Firm 1 sells only A. The best Firm 1 could do is set $P_A = Y_A - \varepsilon$ by Condition 2. Then, profits are $2Y_A - 2\varepsilon - C$, which is less than π^1 because $P_{B*} > \Delta + \tfrac{C}{2}$.

Finally, consider deviations in which Firm 1 sells only B. The best Firm 1 could do is $P_B < P_{B*} - \Delta - \varepsilon$ set by Condition 3. Then profits are $2P_{B*} - 2\Delta - 2\varepsilon - C$, which is less than π^1 by Conditions 1 and 2.

Hence, Firm 1 has no profitable deviation against P_{B*}.

Against $\hat{P}$, does Firm 2 have a profitable deviation? Clearly, Firm 2 does no better for $P > P_{B*}$, since Firm 2 will still sell nothing. So consider deviations $P < P_{B*}$. If $P = P_{B*} - 2\varepsilon$, consumer X will continue to buy the bundle, and consumer Y will switch to B^*. Profits would be $P_{B*} - 2\varepsilon - C < 0$ by $P_{B*} \le C + \varepsilon$. If $P < P_{B*} - (X_A - Y_A) - \varepsilon$, both X and Y will switch to B^*. The highest price that satisfies this constraint is $P = P_{B*} - (X_A - Y_A) - 2\varepsilon$. At this price, profits would be $2P_{B*} - 2(X_A - Y_A) - 4\varepsilon - C$, which is negative for all $P_{B*} \in \left(\Delta + \tfrac{C}{2}, C + \varepsilon \right]$ by Condition 1. This establishes that

$$\left\{ \hat{P} = Y_A - \Delta + P_{B*} - \varepsilon, \ P_{B*} \in \left(\Delta + \tfrac{C}{2}, C + \varepsilon \right] \right\}$$

are Nash equilibria. Q.E.D.

The reader can easily verify that the numerical example given earlier (in which $X_A = 10$, $Y_A = 8$, $X_B = Y_B = 4$, $X_{B*} = Y_{B*} = 5$, $C = 3$) satisfies the conditions of a Nash equilibrium in which Firm 1 excludes Firm 2 from the market via a bundled price of \$9.99 against P_{B*} of \$3.

The bundled pricing equilibria that we derived in this section are not the only equilibria in the market. There are also equilibria in stand-alone prices. These are presented in the following section.

THE GAME WHEN BUNDLING IS PRECLUDED

We now examine the model if bundled pricing is not allowed.

Proposition 2: When bundling is not permitted, all equilibria are in the set

$$\left\{ P_A = Y_A - \varepsilon, \ P_B^u \in \left[\tfrac{C}{2} - \Delta + \varepsilon, \ \tfrac{C}{2} + \varepsilon \right], \ P_{B*}^u = P_B^u + \Delta - \varepsilon \right\}$$

Proof: In any equilibrium in this set, Firm 1 sells A to X and Y, for profits of

$$\pi_R^1 = 2Y_A - 2\varepsilon - C > 0$$

by Condition 3. Firm 2 sells B^* to X and Y for profits of $2(1 - \gamma)\Delta$ for some $\gamma \in [0, 1]$, where

$$P_B^u = \tfrac{C}{2} + \varepsilon - \gamma\Delta.$$

Because the utilities are independent across A and B (by Assumption D), we can examine the prices for B and B^* separately from the price for A. Hence, to verify the equilibrium, first examine P_A. For $P_A < Y_A - \varepsilon$, Firm 1 clearly cannot benefit because it can sell to both customers at $Y_A - \varepsilon$. At $P_A = Y_A$, Customer Y buys A with probability .5, so profits from A are $.5(Y_A) + Y_A - C < \pi_R^1$ (where the subscript R denotes the "regulated" scenario precluding bundling). For $Y_A < P_A < X_A$, Firm 1 sells only to Customer X, for profits of $P_A - C < \pi_R^1$ by Condition 2. For $P_A = X_A$, Customer X buys A with probability .5, and for $P_A > X_A$, no one buys A. Hence, $P_A = Y_A - \varepsilon$ is uniquely optimal.

Now examine the candidate equilibrium pair P_B^u, P_{B*}^u. For any stand-alone price of B, P_B, Firm 2 can win both customers with any $P_{B*} < \Delta + P_B$. The best response by Firm 2 to any P_B is therefore $P_B + \Delta - \varepsilon$ if $2(P_B + \Delta - \varepsilon) \geq C$, and any price greater than or equal to $\Delta + P_B$ otherwise.

Case 1: $P_B + \Delta - \varepsilon \geq \tfrac{C}{2}$. In this case, against $P_{B*} = P_B + \Delta - \varepsilon$, Firm 1 can do strictly better than P_B by offering $P_B - 2\varepsilon$ as long as $P_B - 2\varepsilon > \tfrac{C}{2}$. Firm 1 has no profitable deviation, however, for $P_B \leq \tfrac{C}{2} + \varepsilon$.

Case 2: $P_B + \Delta - \varepsilon < \frac{C}{2}$. In this case, Firm 2's best response to P_B is any $P_{B*} \in (\Delta + P_B, \infty)$. But then Firm 1's profit is $2P_B - C < 0$. Hence, Firm 1 can do better by increasing price to P_{B*}.   Q.E.D.

We can now compare Firm 1's profit in the bundling equilibrium with those in the stand-alone (regulated) equilibrium. In the bundled price equilibrium, recall that profits are

$$\pi^1 = 2Y_A - 2\Delta + 2P_{B*} - 2\varepsilon - 2C, \text{ for } P_{B*} \in \left(\Delta + \frac{C}{2}, C\right].$$

In the stand-alone pricing equilibrium, Firm 1's profits are $\pi_R^1 = 2Y_A - 2\varepsilon - C$. Hence, $\pi^1 > \pi_R^1$ as $P_{B*} > \Delta + \frac{C}{2}$, which always holds in the bundling equilibrium.

Therefore, when a bundled pricing equilibrium exists (Proposition 1), bundled pricing is always preferred by the monopolist to stand-alone pricing.

SOCIAL WELFARE

Total social welfare is profit plus net consumer surplus. In the bundling equilibrium, total social welfare is

$$TSW_B = Y_A + X_A + Y_B + X_B - 2C.$$

Net consumer surplus under bundling is

$$NCS_B = X_A - Y_A + X_{B*} + Y_{B*} - 2P_{B*} + 2\varepsilon.$$

In the regulated stand-alone case

$$TSW_R = Y_A + X_A + Y_B + X_B + 2\Delta - 2C$$
$$\text{and } NCS_R = X_A - Y_A + 2\varepsilon + X_{B*} + Y_{B*} - C - 2(1 - \gamma)\Delta.$$

Hence, $TSW_R > TSW_B$ and $NCS_R > NCS_B$ by $P_{B*} > \Delta + \frac{C}{2}$. That is, welfare, both in terms of social welfare and net consumer surplus, is greater if bundling is precluded. The value of the social gain is the incremental value of the superior product that would be offered if Firm 1 is prevented from bundling, but would not be offered as a result of the exclusionary effect of the bundled pricing.

MIXED BUNDLING AS A REMEDY

One question that naturally arises in antitrust considerations of potentially injurious behavior is whether remedies that can be easily implemented are available. We have shown that one obvious remedy will enhance social welfare and net consumer surplus: namely, prohibiting bundled pricing. A less severe remedy might be to permit bundled pricing but require that stand-alone prices also be offered, which is known as mixed bundling. We examine here whether such a requirement would in fact be an effective remedy.

Clearly, simply requiring that stand-alone prices be offered in addition to the bundled price is not sufficient to preclude anticompetitive behavior because the monopolist could set arbitrarily high stand-alone prices, which would never attract consumers from the bundled offering. It is straightforward to show that a simple adding-up rule is, however, sufficient to overcome the anticompetitive effects of the bundle under the standard bundling scenario we have modeled. We show in the next section, however, that when the model is revised to reflect the new buyer scenario that is relevant to the Microsoft market, this remedy is no longer effective.

Suppose Firm 1 offers stand-alone prices P_A and P_B, as well as a bundled price P_B^b.

Proposition 3: Under the conditions of Proposition 1, there does not exist an equilibrium with mixed bundling such that $P_A + P_B \leq P^b$ (the adding-up rule) and such that Firm 2 is precluded from the market.

Proof: For B^* to be precluded requires either

$$P_B \leq \frac{C}{2} - \Delta, \text{and/or}$$

$$P^b < Y_A - \Delta + P_{B^*} \text{ (by condition (2) of Proposition 1).}$$

Note that Firm 1 will never offer a set of prices that provides profits less than $\pi^A = 2Y_A - 2\varepsilon - C$, because this profit level can be assured by $P_A = Y_A - \varepsilon$ and any P_B that will be rejected by both customers.

Notice that profits on B are always negative for $P_B \leq \frac{C}{2} - \Delta$. Also, the most profits Firm 1 can earn, if $P_B \leq \frac{C}{2} - \Delta$ would be $2Y_A - 2\varepsilon - C + 2P_B - C$, which is inferior to π^A. So Firm 1 would never offer $P_B \leq \frac{C}{2} - \Delta$. For P_B to yield higher profits than π^A requires $P_{B^*} > \Delta + \frac{C}{2} - \varepsilon$. However, if

$$P_{B^*} \geq \Delta + \frac{C}{2} \text{ and } P^b \leq Y_A - \Delta + P_{B^*} - \varepsilon,$$

Firm 2 always has a profitable deviation to $P_{B^*} - \varepsilon$. Hence, there cannot be an equilibrium with $P_B < Y_A - \Delta + P_{B^*}$ in mixed bundling under the adding-up rule. Q.E.D.

Hence, imposing an adding-up rule, by which the monopolist must offer stand-alone prices that sum to no more than the bundled price, is an effective remedy in the standard scenario for ensuring against the exclusionary effect of bundled pricing.

THE NEW BUYER SCENARIO

For certain customers, the software market for operating systems and browsers differs from the modeling approach we have adopted in the following respect. For some consumers, the value of the competitor's browser is zero if the consumer does not buy the operating system, because the consumer does not already own a computer. Consumers who already have a PC can choose to buy an upgrade to the operating system or not, and would buy and use the competitor's browser (and/or the monopolist's browser, if offered separately) even without upgrading to Windows98. Our model fits this case well. For consumers who do not have a PC, however, neither browser has value without an operating system. This case, which we call the new buyer scenario, requires a modified analysis. We show here, as is intuitively apparent, that when consumers cannot use B or B^* without A, the monopolist's position is significantly enhanced, because the price and availability of B^* is no longer a constraint on the price of the bundle. Indeed, the bundling problem is identical to that of a two-goods monopolist and the rival can gain, at most, the incremental value of B^* over B. To capture the new buyer scenario, let:

$$U_i(A, B, B^*) = U_i(A) + Z, \text{ where } Z = \begin{cases} Max\{Ui(B), Ui,(B^*) \text{ if } A > 0\} \\ 0 \qquad\qquad\qquad\qquad \text{ if } A = 0 \end{cases}.$$

Let V_i be consumer i's net consumer surplus function. Assume V_i is continuous in P, where $P = \{P_A, P_B, \hat{P}\}$

Proposition 4: In the new buyer scenario, Firm 1's pricing problem is equivalent to the two-product monopoly problem.

Proof:[9] Firm 1's problem is:

$$\underset{P=\{P_A, P_B, \hat{P}\}}{Max} \pi(P_A, P_B, \hat{P}; C) \text{ such that}$$

$$\min\{V_X(P), V_Y(P)\} = 0.$$

To show that this problem is identical to the two-product monopoly problem, it need only be shown that Condition 5 is the relevant constraint for the new buyer problem. To see that this is the relevant constraint, suppose that at the optimal price vector P^*, $V_i(P^*) > 0$, for $i = X$ and Y. Then Firm 1 could charge $P^* + \varepsilon$, (i.e., each element of the P-vector is increased by ε) for ε sufficiently small, and both consumers would continue to buy from Firm 1, independent of P_{B^*}. In particular, there is no $P_{B^*} \geq 0$ that would induce either customer to buy instead of A or the bundle including A, because the utility value of B^* without A is zero.   Q.E.D.

In the new buyer scenario, the adding-up rule is not a sufficient remedy to preclude anticompetitive bundling, as it was shown to be previously. An example is sufficient to make the point.

---

9. For purposes of the proof we let P vary continuously (i.e., we let ε be arbitrarily small). For finite ε the theorem holds with a slightly more cumbersome proof and Condition 4 holds within an appropriate neighborhood of zero.

Recall the earlier example in which X values A at \$10 and B at \$3, and Y values A at \$8 and B at \$4. A bundling equilibrium that excludes Firm 2 is $\hat{P} = \$9.99$, $P_{B*} = \$3$. Now suppose Firm 1 were required to offer unbundled prices as well, subject to the constraint that $P_A + P_B \leq \hat{P}$. Firm 1 could offer, for example $P_A = \$8.99$, $P_B = \$1$. Because neither customer receives any utility from buying B alone, the best response to these prices for both customers is to buy both A and B from Firm 1. Hence, these prices would satisfy the adding-up rule, but would not divert consumers from the bundle or offer any remedy to the exclusionary effect of bundling.

Not only is the adding-up rule ineffective in the new buyer scenario, but a simple remedy such as prohibiting bundled prices entirely is also insufficient. Again, we can demonstrate this with our example. If Firm 1 did not offer $\hat{P} = \$9.99$, but only offered the "unbundled" prices of $P_A = \$8.99$, $P_B = \$1$, the effect on consumers would be identical. Indeed, Firm 1 could offer prices of $P_A = \$9.99$, $P_B = \$0$, and still sell both products to both customers while making it impossible for Firm 2 to compete.

To prevent the anticompetitive effects of bundling, a stronger remedy is needed. One such viable remedy is the requirement that bundled prices be prohibited and the unbundled prices each cover the average costs of the respective products.[10] To see this, note that under the average cost rule, $P_B \geq \frac{C}{2}$. Hence, Firm 2 could attract both customers and make a profit for any $P_{B*} \in [P_B, P_B + \Delta - \varepsilon]$.

The problem with such a remedy, however, is that it would be very difficult for the regulator to determine average costs and would therefore impose significant and ongoing regulatory infrastructure on the industry, while not necessarily being effective. Structural remedies, such as divestiture and separation of the firm into the A business and the B business, although potentially onerous in other respects, avoid the need for ongoing cost-based regulation.

CONCLUSION

Bundled pricing can be used successfully to leverage a monopoly in one product into a monopoly over another product. The opportunity to bundle is critical to the monopoly leverage in our model, as we demonstrated earlier. The fact that bundling permits a squeeze against a more efficient rival with a superior product makes the outcome particularly costly from a social welfare standpoint, because consumers are worse off if they are denied access to the more desirable product B^*. Moreover, it is a straightforward implication of our model that bundled pricing can serve as a deterrent to innovation. Consider the situation of Firm 2 before it invests in

---

10. This average cost rule differs from the more commonly accepted marginal cost criterion for predatory pricing due to the cost structure assumed in the industry. However, the rules have a common conceptual basis, which is that prices should recover the nonsunk costs of offering the product(s).

designing product B^*. Firm 2 can anticipate that once B^* hits the market, Firm 1 will bundle its products so as to render B^* unprofitable. This discourages Firm 2 from making the investment at all. Even if Firm 2 would enjoy some period of delay before Firm 1 could execute the bundling strategy, it would not be profitable to invest in the innovation as long as the delay is not too great relative to the investment cost. Hence, the conventional wisdom that bundling by Microsoft has dampened innovation is consistent with the results of our model. When bundled pricing is precluded, there is always an equilibrium in which customers can buy their preferred product B^* on an unbundled basis. This maximizes total social welfare. However, in the new buyer scenario, even precluding bundled pricing is not sufficient to alleviate exclusion, because apparently unbundled prices can perfectly mimic the bundled prices. In the latter case, structural remedies may be necessary.

ACKNOWLEDGMENTS

An early version of this analysis was developed as part of the authors' work on behalf of Universal Studios, Inc., an interested party to the deliberations by the Commission of the European Communities regarding the German digital pay TV joint venture proposed by Bertelsmann and Kirch in 1997.

REFERENCES

Adams, W., & Yellen, J. L. (1976). Commodity bundling and the burden of monopoly. *Quarterly Journal of Economics, 90,* 475–498.

Bakos, Y., & Brynjolfsson, E. (1998, October). *Bundling and competition on the Internet: Aggregation strategies for information goods.* Paper presented at the 26th annual Telecommunications Policy Research Conference, Alexandria, VA.

McAfee, R. P., McMillan, J., & Whinston, M. D. (1989). Multiproduct monopoly, commodity bundling, and correlation of values. *Quarterly Journal of Economics, 104,* 371–383.

Owen, B. M., & Wildman, S. S. (1992). *Video economics.* Cambridge, MA: Harvard University Press.

Salinger, M. (1995). A graphical analysis of bundling. *Journal of Business, 68*(1), 85–98.

Stigler, G. J. (1963). United States v. Loew's, Inc.: A note on block booking. In P. Kurland (Ed.), *The Supreme Court review: 1963* (pp. 152–157). Chicago: University of Chicago Press.

Whinston, M. D. (1990). Tying, foreclosure, and exclusion. *American Economic Review, 80*(4), 837–859.

Combination Bidding in Spectrum Auctions

Paul Milgrom
Stanford University

Complementarities among the licenses sold in spectrum auctions offer significant challenges for auction design. When licenses are mutual substitutes, even if bidding is for individual licenses only, participants who bid straightforwardly for the licenses they want do not run the risk of acquiring an unwanted collection. This encourages vigorous competition in the auction and contributes to efficiency in the allocation. Complementarities arising from scale economies seem to have blocked vigorous competition in a spectrum auction in the Netherlands in 1998. Bidders with such scale economies and others with narrower requirements can both be accommodated by an auction design that permits bids for packages of licenses. The considerations in designing such an auction are reviewed.

INTRODUCTION

The U.S. government first began auctioning portions of the radio spectrum for commercial use in 1994. The auction era replaced one in which administrative hearings and even lotteries had been used to assign spectrum. The auction policy was widely hailed for leading to quicker, more market-driven decisions about license assignments, for reducing opportunities for political interference, and for generating substantial revenues for the government.

The standard auction format of the Federal Communication Commission (FCC), called variously the *simultaneous multiple round auction* or the *simultaneous ascending auction*, has also been widely applauded[1] and imitated. Since its

---

1. For example, Cramton (1995, 1997) and McAfee and McMillan (1996) gave favorable accounts of the performance of the auction in the United States.

introduction in the United States, some variant of the simultaneous ascending auction has been announced or used for sales of radio spectrum in Canada, Mexico, Australia, New Zealand, Germany, and the Netherlands. The new auction rules have also been adapted for other kinds of sales. Variants have been used for everything from selling undeveloped housing lots at Stanford University to selling rights for standard offer electrical service in New England. (More detailed descriptions of several of the applications of the new auction can be found at http://www.market-design.com/projects.html)

Despite the successes of the new auction rules relative to older designs, the possibility of doing still better has received increasing attention. One possible improvement that has attracted the attention of both economic theorists and economic experimenters is allowing participants to bid for combinations or "packages" of licenses, rather than just individual licenses. This chapter reports on the main pros and cons of that suggestion and some of the hurdles that must be cleared to implement such a suggestion effectively.

The remainder of this chapter is organized in three sections. The first of these gives a brief description of the simultaneous ascending auction and its rules.

After the rules are described, the following section discusses the theory and evidence about when the auction might need to be adapted to allow bidding for packages of spectrum licenses, rather than just for individual licenses. According to theory, an auction in which bidding is only for individual licenses can be quite effective when the spectrum licenses for sale are close substitutes, but may fail badly when some of the licenses are complements. I argue that evidence from the Netherlands spectrum auction tends to confirm the theoretical prediction. Thus, the theory suggests that the choice of scope and scale of the licenses to be sold is an important element in the design of a successful simultaneous ascending auction because it determines the extent to which the items for sale complement each other.

The last section discusses the two main approaches to bidding for packages: the generalized Vickrey auction and a modified simultaneous ascending auction. The section identifies certain theoretical advantages of the generalized Vickrey auction, but it also identifies several important practical limitations. The alternative dynamical auction has been less completely studied. It lacks the theoretical optimality properties of the Vickrey auction but has certain offsetting practical advantages. The question of how best to structure the rules of the simultaneous auction with package bidding remains an important open question.

DESCRIPTION OF THE SIMULTANEOUS ASCENDING AUCTION

A simultaneous ascending auction is an auction for multiple items in which bidding occurs in rounds. At each round, bidders simultaneously make sealed bids for any licenses in which they are interested. After the bidding, round results are posted. For each item, these results consist of the identities of the new bids and bidders and the standing high bid and the corresponding bidder. The initial standing high bid for each item is zero and the corresponding bidder is the auctioneer. As the auction progresses, the new standing high bid at the end of a round for an item is the larger of the previous standing high bid or the highest new bid and the corresponding bidder is the one who made that bid. In addition to the round results, the minimum bids for the next round are also posted. These are computed from the standing high bid by adding a predetermined bid increment, such as 5% or 10%.

Bidder activity during the auction is controlled by what the FCC called the *Milgrom-Wilson activity rule*. It works as follows. First, a quantity measure for spectrum is established, which provides a rough index of the value of the license. Typically, the quantity measure for a spectrum license is based on the bandwidth of the licensed spectrum and the population of the geographic area covered by the license. At the outset of the auction, each bidder establishes its initial eligibility for bidding by making deposits covering a certain quantity of spectrum. During the auction, a bidder is considered active for a license at a round if it makes an eligible new bid for the license or if it owns the standing high bid from the previous round. At each round, a bidder's activity is constrained not to exceed its eligibility. If a bid is submitted that exceeds the bidder's eligibility, the bid is simply rejected.

According to the original rules, the auction is conducted in three stages. In the first stage, a bidder who wishes to maintain its eligibility must be active on licenses covering some fraction f_1 of its eligibility. If a bidder with eligibility x is active on $y < f_1 x$ during this stage, then its eligibility is reduced at the next round to y/f_1. In the second and third stages, a similar rule applies but using fractions f_2 and f_3. In recent auctions in the United States, the fractions used have been $(f_1, f_2, f_3) = (.6, .8, .95)$. Thus, in Stage 3, bidders know that the auction is nearing its close in the sense that the remaining demand for licenses is just $1/f_3$ times the current activity level.

There are several different options for rules to close the bidding and the spectrum regulator is presently reevaluating these. One proposal, made by McAfee, specified that when a license had received no new bids for a fixed number of rounds, bidding on that license would close. That proposal was coupled with a suggestion that the bid increments for licenses should reflect the bidding activity on a license. A second proposal that Wilson and I made specified that bidding on all licenses should close simultaneously when there is no new bidding on any license. To date, the latter rule is the only one that has been used in the spectrum auctions.

When the auction closes, the licenses are sold at prices equal to the standing high bids to the corresponding bidders. The rules that govern deposits, payment terms, and so on are quite important to the success of the auction,[2] but they are

---

2. Failure to establish these rules properly led to billions of dollars of bidder defaults in the United States "C-block auction." Similar problems on a smaller scale occurred in some Australian spectrum auctions.

mostly separable from the other auction rule issues and receive no further comment here.

WHEN ARE COMBINATION BIDS NEEDED?

One important feature of the auction rules is that bids are made on each license individually. In practice, however, a license is not evaluated in isolation. Rather, a license is used as part of a business plan and its value depends on which other licenses the bidder acquires. Even the identities of other bidders who acquire licenses may affect the value, as that may affect both the likelihood of concluding successful roaming agreements, the technologies that are developed by suppliers, and the nature of competition in the telecommunications market.

This analysis focuses mostly on the value dependency among licenses for a single bidder. Logically, when there are more than two interdependent licenses, the value relations among them can take a large variety of formats. For reasons of simplicity, however, theoretical economic analyses of auctions mainly focus on two special cases. In the first, all of the licenses are substitutes; in the second, there are just two licenses and they are complements. A key question in each case is whether the outcome would be improved if bids for packages were allowed.

According to one theory (Milgrom, 1997), if bidders bid in a straightforward manner for the items they wish to acquire and if the items being sold are economic substitutes, then the simultaneous ascending auction with sufficiently small bid increments will lead to an efficient outcome. Straightforward bidding means that, at each round, the bidder selects its bids to maximize its profits based on the hypothesis that the other bidders will not bid again and the auction will end after the present round.

The intuition behind this conclusion is as follows. Suppose a bidder finds all the licenses sold in the auction to be substitutes, meaning that raising the price of some of the licenses would not make that bidder less eager to buy the others. During the course of the auction, the rising prices of the licenses never make a bidder less willing to acquire the items on which it still has the standing high bid. Consequently, if the bidder bids straightforwardly, then when the auction ends the bidder will never wish to withdraw any bids: It will be satisfied to pay the prices it has bid to acquire its items. Moreover, because it did not bid at the last round, the bidder does not wish to acquire any other item at the final bid price (plus one increment, but an additional argument establishes that if the bid increment is sufficiently small, it never affects the bidder's choice). Hence, the prices and license assignment determined by the auction form a competitive equilibrium outcome. By the First Theorem of Welfare Economics, that implies that the assignment is efficient.

The conclusion is that package bids are not needed to improve the potential outcome of the auction in the event that licenses are substitutes. Package bids may still be desirable for the incentives they provide, which generally depend on the other rules of the auction. For example, the generalized Vickrey auction provides strategic incentives for straightforward bidding but relies on package bidding and price discrimination to achieve that. I set that issue aside for the discussion here.

Although straightforward bidding for individual licenses leads to efficient, competitive equilibrium outcomes when licenses are all mutual substitutes, the situation changes dramatically if some of the licenses are complements, for two reasons. First, as shown in Milgrom (1997), allowing for complements in addition to the substitutes implies that a competitive equilibrium may fail to exist. This conclusion is particularly devastating for the view that an auction is a means to discover equilibrium prices.

Second, the presence of complements enormously complicates a bidder's problem. When licenses are mutual substitutes, a straightforward bidder will always wind up with a set of licenses that it is satisfied to have at the final bid prices. However, if the licenses are complements, a straightforward bidder is exposed to the possibility that it may not want the licenses assigned to it at the prices it must pay. To see this, consider a bidder whose value for the pair of licenses AB is 10 but whose values for A and B individually are just 2 apiece. If the price of each license reaches 3, straightforward bidding means that the bidder should bid 4 for each license, because it wants the pair at those prices. However, if there is no further bidding for License A but the price of License B continues to rise to 7, the bidder is stuck. It does not want to pay the going prices for A and B, but it also does not want to acquire License A alone for a price of 4. From a bidder's perspective, this "exposure problem" makes strategic bidding complicated.

One possible response to the exposure problem is for the bidder to bid cautiously, limiting its exposure to losses from acquiring an unwanted license or licenses. This response is particularly appealing when several licenses must be acquired to obtain good value.

The auction completed in the Netherlands in February 1998 for DCS-1800 spectrum (spectrum in the 1800 MHz assigned for use in "digital communications services") provides a good example of how costly the exposure problem can be. In that auction, complementarities among licenses resulted from economies of scale. Two of the lots on offer, A and B, contained sufficient bands to be used on a stand-alone basis for an efficiently scaled PCS telephone system. Sixteen smaller lots were also offered, but one needed to acquire at least five and perhaps even six or seven of these to build an economically viable business. These numbered lots, however, were useful as spectrum increments to incumbent cellular telephone providers.

The auction outcome reflected this scale problem in a natural way. For comparability, we express the prices of all the lots on a "per band" basis. The prices of the A and B lots were 8.0 million Dutch guilders (NLG) and 7.3 million NLG, respectively. The prices of the numbered lots were much lower, ranging from 2.9

million to 3.6 million NLG per band. Evidently, those who wish to establish new wireless telephone businesses found it too risky even to try to assemble these by bidding individually for the smaller lots.

The practical lesson of the Netherlands experience is that there are real limits on the use of the simultaneous ascending auction. If bidding is to take place on the licenses individually, then the licenses themselves should be structured to keep the exposure problem manageable. If that is not possible, the solution may lie in some form of bidding for packages.

PACKAGE BIDDING SCHEMES

Package bidding schemes can be devised that, in theory, both encourage bidders to bid in a straightforward way (overcoming strategic incentives to misrepresent values or reduce demand in an attempt to manipulate prices) and solve the problem of allocating combinations of licenses. There are two main approaches to this: the generalized Vickrey auction and dynamical package bidding.

The generalized Vickrey method is a complete solution to the package bidding problem for a certain ideal theoretical environment. This involves a generalization of Vickrey's (1961) auction design using elements introduced by Clarke (1971) and Groves (1973). In the relevant environment, each bidder has a valuation for each possible package of licenses and knows those valuations. It reports valuations to the auctioneer, who then computes the total-value maximizing license assignment and implements it. The auctioneer then sets prices so that each bidder's profit, if it reports truthfully, is equal to the amount that the total value is increased by its participation. It can be shown that with this pricing rule, it is in each bidder's interest to report its values truthfully, regardless of how others report. Of course, if each bidder does report truthfully, the result is an efficient license assignment.

A report by Market Design, Inc and Charles River Associates (1997) reviews four practical difficulties of implementing the generalized Vickrey auction in detail. I add a fifth to the following list.

- *Political and legal limitations.* The Vickrey auction makes the profits of the bidders obvious to outside observers, exposing the outcomes to subsequent challenge. In addition, it sometimes applies price discrimination in favor of larger bidders to offset their incentive to reduce prices by withholding demand. Such price discrimination may be illegal or politically impractical.
- *Budget limitations.* The theoretical analysis assumes that the bidders' budgets are unlimited. Some of the important consequences of budget limitations are analyzed in Che and Gale (1996, 1998).
- *Complexity.* Spectrum auctions vary in size, but some involve the sale of hundreds or thousands of licenses. In its pure form, the generalized Vickrey auction entails reports of value for every subset, which is far too many

for practicality. To be practical, some variation would need to be implemented that either reduced the number of combinations to be valued or that limited the valuation formula, so that it can be described with a manageable number of parameters.

- *Common value issues.* A central assumption of the generalized Vickrey analysis is that each bidder knows its own values and is uninterested in the values of others. In practice, however, bidders may well respect each other's expertise and want to weight the other bidders' assessments about demand growth, future technological developments, and so forth. These common value issues can, if important, have drastic implications about the efficacy of alternative auction designs (Klemperer, 1998). Generally, these "common value" issues work against the effectiveness of Vickrey auctions.

- *Dependencies among license winners.* Another assumption of the generalized Vickrey auction is that values depend only on the set of licenses won. As described earlier, however, a bidder may care who the other winners are, for example because that affects the possibility of a roaming agreement. Although it is possible in theory to structure a further generalized Vickrey auction so that prices depend on the entire license assignment, that exacerbates the other difficulties already described.

Partly because of the complexity problem already discussed, some pundits have proposed using an auction that allows the bidders to make both individual license bids and package bids and to specify potentially relevant packages dynamically during the auction. There is evidence from laboratory experiments that such rules may sometimes perform extraordinarily well (Ledyard, Noussair, & Porter, 1994). Theoretical analysis of the experimental auction rules suggests that strategic incentives should be a more serious problem than the experimenters seemed to find (Milgrom, 1997). The central problem is a free-rider problem among bidders for individual licenses who must implicitly form a team to outbid the "package" bidder. The theory holds that even if there are no actual complementarities, a large bidder can sometimes make a package bid that exploits the free-rider problem and allows it to win too many licenses at a low price.

These theoretical predictions are preliminary, and they are also testable in economic laboratory experiments. For these reasons, this is a particularly promising area for both theoretical and experimental research. Even now, such research is proceeding. While this volume was in press, a new design for combinatorial bidding was proposed by DeMartini, Kwasnica, Ledyard, and Porter (1998). Their design is resistant to free-rider problems, and other new designs are being tested that have better theoretical performance if bidders behave naively. The next stages include more laboratory testing and the introduction of these methods to sell items of substantial value. The adoption of these techniques for real auctions is almost certainly imminent.

REFERENCES

Che, Y.-K., & Gale, I. (1996). Financial constraints in auctions: Effects and anti-dotes. *Advances in Applied Microeconomics, 6*, 97–120.

Che, Y.-K., & Gale, I. (1998). Standard auctions with financially constrained bid-ders. *Review of Economic Studies*, 65, 1–21.

Clarke, E. (1971). Multipart pricing of public goods. *Public Choice, 8*, 19–33.

Cramton, P. (1995). Money out of thin air: The nationwide narrowband PCS auc-tion. *Journal of Economics and Management Strategy, 4*, 267–343.

Cramton, P. (1997). The FCC spectrum auctions: An early assessment. *Journal of Economics and Management Strategy, 6*(3), 431–495.

DeMartini, C., Kwasnica, A., Ledyard, J., & Porter, D. (1999). *A new and im-proved design for multi-object, iterative auctions* Pasadena, CA: Cal Tech Working Paper. Available at http://www.hss.caltech.edu/~akwas/rad.pdf.

Groves, T. (1973). Incentives in teams. *Econometrica, 41*, 617–631.

Klemperer, P. (1998). Auctions with almost common values: The "wallet game" and its applications. *European Economic Review*, 42, 757–769.

Ledyard, J., Noussair, C., & Porter, D. (1994). *The allocation of shared resources within an organization* (IRBEMS Working Paper No. 1063). West Lafayette, Indiana: Purdue University.

Market Design, Inc, & Charles River Associates. (1997). *Report 1B: Package bid-ding for spectrum licenses* (Report to the U.S. Federal Communications Com-mission). Available http://www.market~design.com/projects.html

McAfee, R. P., & McMillan, J. (1996). Analyzing the airwaves auctions. *Journal of Economics Perspectives, 10*, 159–175.

Milgrom, P. (1997). Putting auction theory to work: The simultaneous ascending auction (working paper, Department of Economics) Staford, CA: Stanford Uni-versity. Available at http://www-econ.stanford.edu/econ/workp/swp98002.pdf

Vickrey, W. (1961). Counterspeculation, auctions, and competitive sealed tenders. *Journal of Finance, 16*, 8–37.

Regulating Anticompetitive Behavior in the Internet Market: An Applied Imputation Model for Peru

Arturo Briceño
OSIPTEL

In some developing countries like Peru, the development of the telecommunications sector is based on promoting vertical integration of the incumbent operator, which is allowed to provide basic telephone services as well as value-added services. After privatization took place in 1992, a legal monopoly for basic telephone services was permitted until 1998, during which time the incumbent could also enter into competitive services such as provision of Internet access. This chapter presents one of the analytical tools that the Peruvian regulator used in a legal dispute between the largest independent Internet service provider (ISP) and the dominant vertically integrated firm. The ISP accused the dominant firm of anticompetitive behavior in the form of price discrimination in the provision of access to essential facilities that any ISP has to employ for the provision of Internet services. The tool was to use an imputation test to determine whether the vertically integrated firm was charging the independent ISP the same prices it was charging itself and its affiliate firms for essential inputs such as telephone lines and dedicated circuits or leased lines. For that purpose a bottom-up model of an efficient ISP was developed, assuming the best technology available in the country for providing dial-up and dedicated access to Internet services. Based on the model, an average incremental cost for providing Internet services was estimated, giving the regulator a good proxy of a price floor that Internet services should have in the market. Any price below the floor may be an indication of price discrimination or the potential presence of cross-subsidies from the noncompetitive services to the Internet services.

INTRODUCTION

In Peru, the legal framework for telecommunications promotes vertical integration among firms. Ever since state-owned telecommunications enterprises were privatized and merged in 1994, there has existed a vertically integrated incumbent firm, Telefónica del Perú (TdP), that is allowed to enter any telecommunications market, including the provision of Internet access services. In addition, as an important ingredient of the privatization process, TdP was awarded a monopoly franchise until 1998 over a set of basic telecommunications services, including local and long-distance telephone service and dedicated circuits. Prices of basic telephone services for the period 1994 through 1998 were set up in advance of the privatization bidding, so that the winning bidder would be obliged to comply with the pre-established tariff schedules in a process known as the *rebalancing program.* Prices of dedicated circuits and other services were also regulated through price ceilings imposed by the regulatory authority, the Supervisory Office for Private Investment in Telecommunications (OSIPTEL).

In this context, established ISPs had to build their backbones by leasing all their transmission capacity from TdP, including dedicated circuits, telephone lines, hunting lines, and so forth. When TdP got into the Internet market in 1997, the largest ISP at that time filed a complaint against TdP. The allegation was that TdP was acting in an anticompetitive manner because, among other reasons, it was using part of its telephone infrastructure but not imputing to itself the same price charged to its competitors.[1] To assess the allegation of discriminatory pricing in inputs used to provide Internet service, OSIPTEL implemented an imputation test to determine whether TdP was charging to the independent ISP the same prices that it was charging itself and its affiliate firms for noncompetitive inputs such as telephone lines and dedicated circuits. For that purpose, a bottom-up model for an efficient ISP was built, assuming the best technology available in the country at that time to provide dial-up and dedicated access to Internet services. Based on the model, the average incremental cost for providing Internet service was estimated, giving a good proxy for the price floor that it should have in the market. Any price below the floor may be an indication of price discrimination or the potential presence of cross-subsidization from the noncompetitive services to Internet services.

The remainder of the chapter is organized as follows. The next two sections give an overview of the Internet market and the key features of the Internet service provisions in 1996 and 1997, from which the anticompetitive allegation emerged. After that, I set up the regulatory problem that this chapter tries to ad-

---

1. The complete set of allegations included not only anticompetitive pricing (cross-subsidization, undue discount practices, etc.) but also nonpricing issues (refusal to supply or connect services, unreasonable delay in supply, etc.). In this chapter I concentrate only on the issue of discriminatory pricing in the use of inputs to provide Internet services.

dress. I then present the cost model used for imputation purposes to determine the degree of price discrimination in the provision of essential services. Next, I present a variant of the model that seeks to mimic the costs that a TdP affiliate would incur to serve the same market as that of an independent ISP. Finally, I present some concluding remarks.

OVERVIEW OF THE PERUVIAN INTERNET MARKET

Internet service in Peru started in 1991. In 1996, before TdP entered the Internet market, there were two other ISPs offering Internet access: Red Científica Peruana (RCP) and International Business Machines (IBM). Both firms provided Internet access services through dial-up (using a modem plugged to an ordinary telephone line) and dedicated links (using dedicated circuits), but due to the legal temporary monopoly awarded to TdP, they were not allowed to deploy their own transmission links or local loops, so these inputs had to be leased from TdP.

The Peruvian Internet market may be considered small by international standards. In 1996, there were approximately 2,700 dial-up Internet users and more than 100 dedicated users (see Table 3.1), which were largely businesses and institutions linked to an ISP with low-speed dedicated circuits. Since then, the Internet market has achieved impressive rates of growth. The monthly rate of growth in the dial-up segment was 8% between 1996 and 1998. A similar growth rate has been reported for the dedicated users. TdP's market shares have increased substantially in this short period. Thus, the TdP market share in the dial-up segment has reached 36%, and its share in the dedicated segment is 57%.

TABLE 3.1
Peru's Internet Market by Service Providers

	1996	1998
Dial-up (users)	2,696	16,000
RCP	2,071	7,800
TdP's affiliates	0	5,700
IBM	625	2,500
Dedicated (number of circuits)	102	560
RCP	78	190
TdP	0	320
IBM	24	50

Source: RCP and author's estimates

TABLE 3.2
Internet Access by PSTN (Standard PSTN Rates),
US$ Purchasing Power Parity

	Internet Bill [1]	Telephone Bill (a)	Telephone Bill (b)	[2]=(a)+(b)	Internet and Telephone [3]=[1]+[2]	Internet/ Telephone [4]=[1]/[2]
Peru (December 1997)						
TdP's IAP (average)	12	17	29	47	58	0.25
RCP	18	17	29	47	65	0.39
Rest of world (January 1995)						
Competitive countries*	27	13	22	35	63	0.78
OECD countries	66	13	29	43	108	1.54

Notes:
[1] Equal to monthly rental plus 1/36 of installation tariff.
(a) Fixed charge per month, consisting of a residential rental plus 1/60 of the installation tariff.
(b) Usage charge, based on 20 calls with a duration of one hour at standard rates.
* U.S., Canada, New Zealand, Sweden, Australia, U.K., Japan.

Source: For Peru, author's elaboration based on observed market tariffs. For rest of world: OECD (1996).

In terms of Internet tariffs and the total bill that a dial-up subscriber has to pay to get access through a public switched telephone network (PSTN), there are important features to be noted by comparing Peru with two groups of countries: the most competitive and all members of the Organization for Economic Co-Operation and Development (OECD).

- The first feature, from column 1 in Table 3.2, is that Internet tariffs in Peru are among the lowest in the world. They are even lower than those in competitive or OECD countries.
- In Peru the lowest Internet tariffs are charged by TdP's Internet access providers (IAPs). Thus, at the end of 1997, TdP's IAPs average tariffs were $12 (U.S.) a month, and RCP was charging $18 (U.S.) a month.
- Telephone bills are more expensive in Peru than in competitive or OECD countries (see column 2).
- Adding up the Internet bill with the telephone bill that a dial-up subscriber has to pay, the total bill in Peru is the same as in competitive countries and lower than in OECD countries (see column 3).

The Internet/telephone relative price (i.e., Internet bill/telephone bill) is much lower in Peru than in other countries (see column 4). The pricing scheme would act as a two-part tariff: The fixed tariff would be low but the usage tariff would be high at international standards. As a result, there could be more Internet subscribers who would each consume less. The overriding factor, however, is that the total

Internet fees are low by international standards and that Internet subscribers have high usage rates of telephone lines.

FIGURE 3.1
TdP and its IAP.

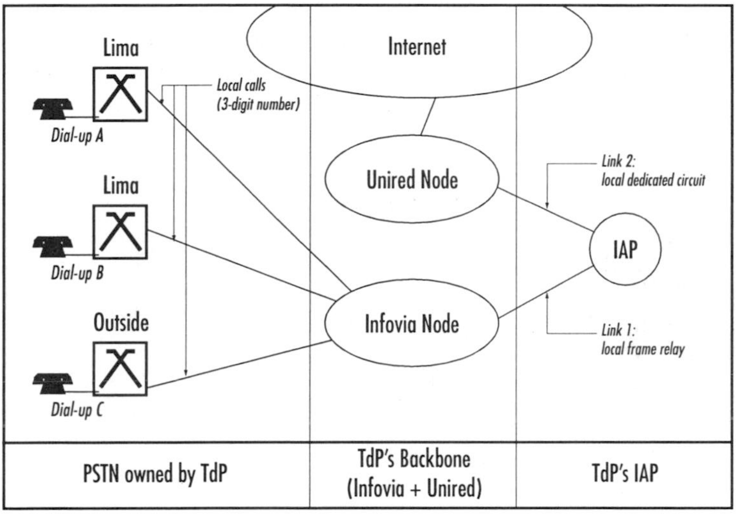

FEATURES OF THE INTERNET SERVICE PROVISION IN PERU, 1996 AND 1997

There are two main ways to access the Internet: (a) through a dial-up or modem, using an ordinary telephone line connected to an ISP or IAP, plus a software package; and (b) through a dedicated link giving a permanent access to an ISP. In this section, I highlight the key features of the Internet services provided by TdP, its affiliates, and an independent ISP.

TdP's Internet Service Provision

In 1997, TdP launched simultaneously two related services: Infovia and Unired.

- Infovia: Any person with a telephone line, a modem, and a free-of-charge software package provided by TdP may access an intranet by dialing a three-digit number (155). The only charge that the telephone user has to pay is the local measured telephone charge, regardless of whether the user is located in or outside Lima (see Fig. 3.1). However, those who want to access the Internet have to subscribe to the service through one of TdP's affiliates or TdP's IAP, which were created for this purpose. TdP's IAPs have access to Infovia by leasing a frame relay circuit from TdP.[2] Note that for any of TdP's IAPs to have national coverage, it needs only a local link to the node closer to Infovia. If the demand of dial-up Internet access increases, TdP's IAP will only have to request a wider local circuit band.

- Unired: This is TdP's dedicated Internet access service. It is provided directly by TdP, which competes in this way not only with the independent ISP but also with its related IAPs. An IAP of TdP has to lease a local dedicated circuit from TdP to have access to the Internet. TdP charges the IAP for two services: (a) the dedicated circuit service and (b) access to Internet service (i.e., Unired).[3]

FIGURE 3.2
Independent ISP.

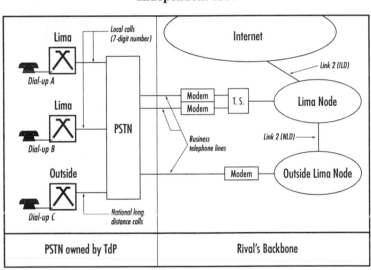

The Independent's Internet Service Provision

- To have Internet access, an independent ISP's dial-up user has to place a telephone call to the closest node of the ISP. For example, a user in a province outside Lima in which there is no physical presence of the ISP (e.g., dial-up user C in Fig. 3.2) has to place a long-distance call to the closest city in which the ISP has a node.

---

2. Currently there are three to four TdP IAPs that capture more than 80% of TdP's dial-up market. TdP's IAPs buy Infovia and Unired services from TdP, but do not have their own infrastructure, instead using the one provided by Infovia (telephone lines, modems, etc.).

3. Access to Unired can take place through dedicated circuits, frame relay, or X.25.

- An independent ISP needs commercial telephone lines in each node to receive phone calls, modems in each of these lines, terminal servers (T.S. in Fig. 3.2), and transmission links to connect its nodes (in the case of Fig. 3.2, a national long-distance link). Facing an increase in demand, an independent ISP must lease more lines, buy additional modems, and, if necessary, establish a new node.
- Besides the transmission links that it must lease from TdP to connect its domestic nodes, an independent ISP must lease an international half long-distance circuit to route its traffic overseas, plus incur the cost of placing its traffic into an international network of another country (e.g., the United States).

Before we turn to the regulatory problem, it is important to point out what I discuss later in this chapter. I present a cost model for a firm like the one depicted in Fig. 3.2. Based on that, I estimate the total service long-run incremental cost for providing Internet services, assumed to be the true incremental cost that any ISP must incur to provide services. Then I impute that estimated cost to the TdP official cost figure to compare them and determine whether there is a revenue shortfall resulting either from price discrimination or economies of scope in the provision of essential services. Later I present a variant of the baseline cost model that attempts to estimate costs for an IAP of TdP similar to that depicted by the righthand panel of Fig. 3.1.

SETTING UP THE REGULATORY PROBLEM

To provide Internet access with national coverage, an ISP needs to have a transmission network or backbone, which, in turn, uses two types of complementary infrastructure: noncompetitive inputs[4] and competitive inputs. *Noncompetitive inputs* consist of the transmission infrastructure required to provide the Internet access, such as dedicated circuits (switched and nonswitched) and telephone lines. Due to its monopoly position over these services, these transmission facilities have to be rented from TdP. The *competitive inputs* correspond to the set of infrastructure and other inputs needed to produce Internet access service that are subject to free and effective competition, such as routers, terminal servers, modems, computers, labor, and so on.

Competitive problems may arise because TdP controls and owns noncompetitive inputs, which are inputs to be sold to itself and rival firms to produce a downstream competitive service (Internet access). RCP's basic allegation was that TdP's dominant position in basic telecommunications markets allowed it to compete unfairly by extending market power into the Internet access market. That kind of undue behavior by TdP may entail, for instance, attempts to raise rivals' costs, by which TdP

---

4. Other names used in the specialized literature are *essential facilities* or *bottleneck inputs*.

would seek to increase prices of essential inputs sold to rivals, lowering the price of the final service even if it also hurts its downstream affiliates. Therefore, as a whole, TdP is maximizing profits through cross-subsidizing services.

To assess the allegation of discriminatory pricing of inputs used to provide Internet service, an imputation test was used to assess whether TdP was charging independent ISPs the same prices for noncompetitive inputs, such as telephone lines and dedicated circuits, that it was charging itself and its affiliate firms. For that purpose, a bottom-up model of an efficient ISP was built, assuming the best technology available in the country for providing dial-up and dedicated access to Internet services. Based on the model, the average incremental cost for providing Internet services was calculated, giving a good proxy for the price floor that Internet services should have in the market. Any price below the floor would be an indication of price discrimination or the potential presence of cross-subsidization from the noncompetitive services to Internet services. By the neutrality and non-discrimination principles contemplated in the Telecommunications Law, TdP should charge itself the same tariffs that it charges its competitors. The determination whether the prices imputed to itself by TdP were the same that it charged to its competitors was called the *imputation test.*[5]

FIGURE 3.3
A schematic view of an ISP backbone.

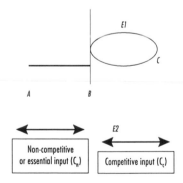

In Fig. 3.3, the technology for producing Internet access service is depicted in a simple and schematic way. The monopolist, TdP, owns a set of noncompetitive in-

5. The application of the imputation concept is similar to the determination of the "transfer price" in the sale of goods and services transferred between units or business centers in a determined given firm. For an exposition of imputation policies in the context of telecommunications regulation, see Larson and Parsons (1993).

puts, represented by the segment AB, needed to produce the "through service" called Internet access service. Any ISP must use the noncompetitive inputs, which are provided only by TdP. The same firm, TdP, jointly with its IAPs, also provides service from Point B to Point C. The competitive inputs used by TdP and its IAPs are represented by the segment B–E1–C. Alternatively, the competitive inputs for an independent ISP are represented by the segment B–E2–C.

Given that the production process demands two sets of inputs, production costs may be grouped into two categories: costs of noncompetitive inputs, which mainly correspond to transmission costs, called C_n; and costs of competitive inputs, which include the rest of the costs related to the provision of the service, denominated C_c. Given a revenue P for the access service to Internet, the following intertemporal restriction (i.e., during the useful life of the service) should hold:

$$P \geq C_n + C_c \tag{1}$$

That is to say that the revenue from the service should cover at least production costs. Data for the lefthand side were easy to obtain because they were in the public domain. The problem was to get reasonable numbers for the righthand variables. Tariffs for noncompetitive inputs (dedicated circuits, telephone lines) were also publicly available as OSIPTEL regulates tariffs for these services.[6] We still needed to have quantities of inputs used. TdP presented to OSIPTEL cost information of its Infovia and Unired services as part of tariff approval proceedings, but compelling reasons made it necessary to contrast that with information obtained from independent calculations. Thus, it was decided to build a bottom-up cost model for a hypothetical ISP firm to estimate the incremental costs of producing the access service to the Internet.

Once production costs were obtained and broken down into noncompetitive and competitive costs, these values were compared and then imputed to the cost figures presented by TdP and checked to see if restriction (1) held.[7] Thus total cost gave an estimate of the floor price for the Internet access service. If after the imputation of the estimated cost to TdP, its revenue is lower than its costs, then discriminatory treatment in access to essential inputs would be indicated. This

---

6. The use of tariffs instead of costs is explained because the provision of access services to Internet would be using network services that could be dedicated to provide other services that themselves could generate additional income for TdP.

7. The assumption behind the imputation tests in this case is that the hypothetical entering firm is at least as efficient as TdP both in the essential resource stretch and in the competitive resource.

might indicate the existence of cross-subsidization from regulated services to competitive services.[8]

A COST MODEL FOR AN INDEPENDENT ISP

In this section, I present the methodological steps followed to build a cost model for an access and backbone ISP that provides basic Web browsing services. As mentioned before, the model estimates the long-run incremental costs that an entrant would incur to provide Internet access service in the Peruvian market, as an independent provider (i.e., not becoming a TdP affiliate). The increment used in the model should be interpreted as the entire marketable output from providing a new service. The costs have been grouped into two categories: noncompetitive and competitive costs. The former consist of transmission costs that a firm has to incur to provide the final service. These costs have been valued using official tariffs for the respective services. The firm has to incur remaining competitive costs to provide the access service to Internet, such as personnel, terminal servers, routers, hardware and software, and so on.

TABLE 3.3
Demand

	Year 1	Year 2	Year 3	Year 4	Year 5
Subscribers (end of year)					
Dial-up	3,000	9,000	15,000	21,000	27,000
Dedicated	80	200	320	440	560
Consumption (*in thousands of megabytes*)	755	2,102	3,450	4,798	6,146
Dial-up	432	1,296	2,160	3,024	3,888
Dedicated	323	806	1,290	1,774	2,258

8. In the context of the price of the access service to the local network, Vogelsang and Mitchell (1997) discussed the case of a local operator that can provide local access for its competitive service of long distance at a lower price than it charges a long-distance competitor and still not show cross-subsidies. The imputation practice serves to determine the floor price that the local access provider should charge itself. If the local access provider charges itself a price below such a floor, the authors argued that evidence of a cross-subsidy exists from the regulated service—access to the local network—to the competitive service—long distance.

Demand

The first step was to estimate the total market that an ISP would achieve over a 5-year planning horizon. The model was built to consider two types of users: dial-up and dedicated subscribers. For each, a given rate of growth over time was assumed: an annual increase of 6,000 dial-up subscribers and 120 dedicated subscribers. It was further assumed that consumption per user was 12 megabytes a month and that for each dedicated subscriber there were 28 users. Table 3.3 shows the demand profile over the 5 years forecast, both in number of subscribers and megabytes per year.

Topology and Technical Dimensioning of the Network

Once the market was estimated, we proceeded to determine the network configuration for the firm. Figure 3.4 summarizes the technical design employed. It has been assumed that the backbone is composed of four main nodes located in three main cities of Peru: Lima (Central Coast of Peru), Arequipa (Southern Coast), and Trujillo (Northern Coast). With the demand information, the next step was to build a traffic matrix for these nodes and then to estimate the required transmission capacity to carry the traffic out.

FIGURE 3.4
Network topology.

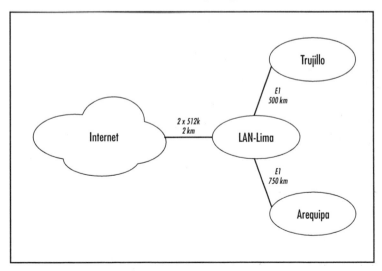

Cost Categories

NONCOMPETITIVE COSTS

As was mentioned before, noncompetitive costs are transmission costs incurred by the firm to transport Internet traffic. Besides dedicated circuits, these costs include commercial telephone lines, which also have to be leased from TdP. The derived demand of transmission inputs is depicted in the bottom part of Table 3.4. Given the evolution of demand, the derived transmission links demand at Year 5 was as follows: 2 E1 for Lima-Arequipa, 1 E1 for Lima-Trujillo, and 10 links of 512 kb (equivalent to almost 3 E1) for the international link.[9] The demand for commercial telephone lines started at 150 lines and increased to 1,350 lines at the end of the project. With the final regulated two-part tariffs ruling at that time for these leased services, we obtained the transmission costs for the firm, which are shown in the upper part of Table 3.5.

TABLE 3.4
Labor and Transmission Input Demands:
Case of an Independent ISP

	Year 0	Year 1	Year 2	Year 3	Year 4	Year 5
Labor demand (persons)		11	16	19	20	23
Professional		4	7	8	8	10
Technical		7	9	11	12	13
Dedicated circuits						
Each period		432	1,296	2,160	3,024	3,888
Lima-Arequipa (E1)		1	0	1	0	0
Lima-Trujillo (E1)		1	0	0	0	0
International (512Kb)		2	2	2	2	2
Commercial lines	150	300	300	300	300	0
Accumulated						
Lima-Arequipa (E1)		1	1	2	2	2
Lima-Trujillo (E1)		1	1	1	1	1
International (512Kb)		2	4	6	8	10
Commericial lines	150	450	750	1,050	1,350	1,350

9. The reason 512 kb dedicated circuits for international transmission were considered, instead of E1, was because until the end of 1996 these were the maximum capacity circuits for international links that TdP provided.

TABLE 3.5
Costs Estimation: Case of an Independent ISP (in thousands of US$)

Noncompetitive costs	Year 0	Year 1	Year 2	Year 3	Year 4	Year 5
Installation tariffs	111	191	196	191	191	7
Dedicated circuits: Lima-Arequipa	5	0	5	0	0	9
Dedicated circuits: Lima-Trujillo	5	0	0	0	0	5
Dedicated circuits: International	12	12	12	12	12	58
Commercial telephone lines	90	179	179	179	179	0
Monthly rentals	0	609	909	1,350	1,649	1,891
Dedicated circuits: Lima-Arequipa	0	140	140	280	280	280
Dedicated circuits: Lima-Trujillo	0	140	140	140	140	140
Dedicated circuits: International	0	242	484	726	967	1,209
Commercial telephone lines	0	87	145	203	261	261
Total noncompetitive costs	111	800	1,105	1,541	1,841	1,964

Competitive costs	Year 0	Year 1	Year 2	Year 3	Year 4	Year 5
Investment	1,767	196	96	96	96	0
Capital equipment	1,567	196	96	96	96	0
Installation	200	0	0	0	0	0
Operating costs	0	378	492	546	558	630
Wages	0	204	318	372	384	456
Rent of building	0	72	72	72	72	72
Energy	0	30	30	30	30	456
Materials	0	72	72	72	72	72
Bad debts (1% of revenues)	0	7	20	32	45	57
Income tax (30% of gross profits)	0	0	83	299	569	821
Labor sharing (10% of operating profit)	0	0	28	90	190	274
Other expenses	0	180	180	180	180	180
Total competitive costs	1,767	761	599	1,243	1,637	1,961
Total cost	1,877	1,562	2,005	2,783	3,478	3,925
Ratio: noncompetitive/total	6%	51%	55%	55%	53%	50%

COMPETITIVE COSTS

- *Capital equipment.* This is an important part of the competitive costs. The detailed list of items is contained in Table 3.6. Overall $2 million (U.S.) was estimated for investment in capital equipment. It comprises routers, terminal servers, modems, software, and so forth.

- *Operating costs.* This includes wages, rent, energy, and materials. The labor demand requirements are presented in the upper part of Table 3.4.
- *Other costs.* This includes categories such as bad debts, income tax, and labor sharing of profits.
- *Other expenses.* These include items such as a supervisory levy[10] and other general and administrative expenses.

Table 3.5 summarizes the total costs of the firm. All of these costs were valued at market prices faced by the firm. As shown, noncompetitive costs account for around 45% of total costs.[11]

Incremental Cost for an Independent ISP

The net present value for the noncompetitive and competitive costs was estimated using a discount rate of 15%. Dividing each category of cost (noncompetitive and competitive, Table 3.5) by the total discounted consumption (Table 3.3), the incremental cost for the Internet access services was obtained. This is shown in column 1 of Table 3.7. The incremental cost of noncompetitive inputs was $0.454 (U.S.) per megabit, whereas the cost of the competitive input was $0.566 (U.S.) per megabit.[12] The floor price to be found was the minimum price that allows the firm just to recover its total costs over the lifetime of the business plan. According to our calculation, that price, before any profit, was $12.2 (U.S.) a month ($1.02 x 12 Mb) for a dial-up subscriber and $343 (U.S.) a month (28 x $12.2) for a dedicated subscriber.

One issue that should be noted concerns the subscriber mix of dial-up and dedicated users assumed. That ratio is assumed to be approximately 36% during the period. An interesting exercise that is beyond the scope of this chapter is to estimate the stand-alone cost of providing just dial-up access, the stand-alone cost of providing just dedicated access, and then to compare these results with the cost of providing both services together. Preliminary results suggest the existence of economies in the joint production of dial-up and dedicated access.[13]

---

10. According to Peruvian law, any telecommunication firm that provides value-added services is subject to an annual payment for OSIPTEL's supervision amounting 0.5% of gross revenues actually billed and received for telecommunications services for each period. In addition, concession holders such as TdP must pay a universal service levy, amounting to 1.0% of gross revenues actually billed and received for telecommunications services for each period after deduction of sales tax.

11. For the United States, Srinagesh (1995) reported that Internet protocol (IP) transport accounts for 25% to 40% of a typical ISPs total costs.

12. All costs are expressed per megabit. Because we have two types of subscribers but no distinct cost information we could not assess incremental costs separately per subscriber and per megabit.

TABLE 3.6
Capital Equipment Estimation: Case of an Independent ISP
(in thousands of US$)

	Unit cost (in US$ '000s)	Year 0 Q	Year 0 Cost	Year 1 Q	Year 1 Cost	Year 2 Q	Year 2 Cost	Year 3 Q	Year 3 Cost	Year 4 Q	Year 4 Cost
Hardware											
Router	35	3	105	0		0		0		0	
Server	35	8	280	0		0		0		0	
Workstation	3	25	75	0		0		0		0	
Switch-hub	5	3	15	0		0		0		0	
Modems*	0.2	150	30	300	60	300	60	300	60	300	60
Terminal servers	6	3	18	6	36	6	36	6	36	6	36
Printers	2	4	8	0		0		0		0	
Scanner	2	5	10	0		0		0		0	
CD-ROM	0.4	5	2	0		0		0		0	
Digital camera	1.5	5	8	0		0		0		0	
Tape backups	1.5	4	6	0		0		0		0	
Instruments	260	1	260	0		0		0		0	
Spare parts	25	1	25	0		0		0		0	
Other	25	1	25	0		0		0		0	
Software	600		600	0		0		0		0	
Solaris 2.4											
CISCO Works											
SunNet Manager											
Graphics											
Programs											
Officemate											
Training	200	0.5	100	0.5	100	0		0		0	
Total			1,567		196		96		96		96
Net Present Value			1,987								

* The assumed ratio of subscribers to modems is 20.
Note: Q is quantity.

13. For instance, the input demands derived from an increase in final demand coming from dedicated subscribers may differ from those coming from dial-up subscribers. In the former case, the firm would only need to implement a new exit in its router (maybe installing an additional card). In the latter case, the firm, to maintain its quality parameters, would need to get new commercial lines, new modems, and terminal servers.

TABLE 3.7
Incremental Cost Estimates: Case of an Independent ISP (US$ per Megabit)

	Bottom-up model	TdP's Infovia	TdP's Unired	Weighted average	Comparison %
	(1)	(2)	(3)	(4)=(2)+(3)	(5)=(4)/(1)
(a) Noncompetitive cost	0.454	0.052	0.193	0.084	19%
(b) Competitive cost	0.566	0.704	0.535	0.665	117%
Total (a) + (b)	1.020	0.755	0.728	0.749	73%

Notes:
(1) Results from the bottom-up cost model. Source: Tables 3, 4, 5, and 6.
(2) Elaborated from the data submitted by TdP.
(3) Elaborated from the data submitted by TdP.
(4) Column 2 (77%) plus Column 3 (23%).

Comparison With TdP's Cost

In 1996 TdP submitted to OSIPTEL revenue and cost information on its Infovia and Unired services.[14] TdP's cost information is summarized in columns 2 to 4 in Table 3.7, under the same format by which costs are grouped in the bottom-up model.

NONCOMPETITIVE COST

As can be seen in column 4, TdP's implicit noncompetitive incremental cost of transmission was $0.084 (U.S.) per megabit, or barely 19% of the cost given by our model. When we replaced the TdP's noncompetitive incremental cost by our own estimation, we found out that TdP was pricing its Infovia and Unired services below the level required to cover their long-run costs. Specifically we estimated that the revenue shortfall was 16%. This was considered a strong indication that TdP was not imputing to itself and its IAPs the same price it did to the rival ISP.

TdP's lower noncompetitive costs could also be explained, at least partially, by economies of scope rather than by underpricing because some resources are shared in the production of TdP's other services. Because noncompetitive costs comprise dedicated circuits and commercial telephone lines, these would be the resources shared by TdP in the provision of other services. TdP's cost information submitted to OSIPTEL contained only one cost item (called "transmission") in the case of Infovia and two cost items (denominated "transmission" and "international link") in the case of Unired. The average incremental noncompetitive costs according to these concepts were $0.052 (U.S.) and $0.192 (U.S.) for Infovia and Unired, respectively. These values may reflect the true

14. TdP submitted this information because it wanted approval of tariffs for bundled services that included nonregulated (e.g., Internet access services) and regulated services (e.g., dedicated circuits).

incremental cost for TdP in the provision of its services and may be lower than the prices for its competitors due to economies of scope. However, our imputation exercise precisely pretends to measure the extent of price differential between what TdP imputes itself and what it charges its competitors. Moreover, we believe that in the Peruvian Internet market, TdP's competitors should benefit from TdP's economies of scope, if they do exist, in the provision of non-competitive inputs.[15] To justify its current pricing, TdP would therefore have to show that it cannot offer the noncompetitive inputs to its competitors without losing the economies of scope.

TABLE 3.8
Comparison of Competitive Cost Items (US$ per Megabit)

	Independent ISP	TdP's Infovia	TdP's Unired
Investment	0.206	0.186	0.164
Capital equipment	0.187	0.186	0.164
Installation	0.019	0.000	0.000
Operating costs	0.164	0.129	0.093
Wages	0.108	0.094	0.068
Rent of building	0.023	0.000	0.000
Energy	0.010	0.007	0.006
Materials	0.023	0.029	0.019
Bad debts (1% of revenues)	0.009	0.014	0.008
Income tax (30% of gross profits)	0.096	0.044	0.105
Labor sharing (10% of operating profit)	0.031	0.119	0.039
Other expenses	0.059	0.212	0.126
Technology and management fee	0.000	0.172	0.105
Other	0.059	0.039	0.021
Total	0.566	0.704	0.535

COMPETITIVE COSTS

Average incremental competitive costs calculated in our model were similar to Unired costs but 20% lower than Infovia costs. Table 3.8 compares competitive costs for the independent ISP and TdP. Investment costs are similar for an independent ISP and TdP, except the installation cost that is not taken into account by TdP. The ISP's operating costs are higher than TdP basically due to rent for build-

---

15. This argument was used by Mitchell and Vogelsang (1998) in their discussion of economies of scope in the context of interconnection pricing between a dominant operator and an interconnecting operator.

ings. The most notable difference among cost categories corresponds to other expenses. This cost is $0.059 (U.S.) per megabit for the independent ISP for items such as supervisory levy and other general and administrative expenses.

However, for TdP's services, the other expenses category also includes a "technology and management fees" item, which amounts to $0.172 (U.S.) out of the total $0.704 (U.S.) competitive cost in the Infovia service and $0.105 (U.S.) out of the total $0.535 (U.S.) competitive cost of Unired. Where does this cost come from? Under the terms of the management contracts signed up by TdP and Telefónica de España, TdP pays Telefónica International, a subsidiary of Telefónica de España, a quarterly "technology transfer fee" of 1% of TdP's operating revenues and a quarterly "management fee" of 9% of TdP's operating profit as set forth in the contracts.[16] These technology and management fees were considered in TdP's cost information, but they are not applicable to an independent ISP.

COSTS FOR AN IAP OF TDP

Figure 3.1 depicted TdP's Internet service provision, which included the role played by IAPs of TdP. Specifically, we showed that to provide access services to the Internet, an IAP of TdP has to lease local links from TdP (righthand side of Fig. 3.1). What would be the incremental cost for an IAP of TdP that served the same market as the independent ISP in the baseline scenario? To address this question, we recalculated the model by varying the estimates of transmission costs and capital investment. The remaining costs were kept as in the baseline scenario (see Tables 3.9, 3.10, and 3.11).

TABLE 3.9
Labor and Transmission Input Demands: Case of TdP's IAP

	Year 0	Year 1	Year 2	Year 3	Year 4	Year 5
Labor demand (persons)		11	16	19	20	23
Professional		4	7	8	8	10
Technicians		7	9	11	12	13
Dedicated circuits						
Each period						
Unired via dedicated circuit (E1)		1	0	1	0	0
Infovia via frame relay (E1)		1	0	1	0	0
Accumulated						
Unired via dedicated circuit (E1)		1	1	2	2	2
Infovia via frame relay (E1)		1	1	2	2	2

16. Technology transfer and management fees are subject to taxes imposed by the Peruvian government on Telefónica Internacional. The amount of such taxes varies between 10% and 30%.

TABLE 3.10
Capital Equipment Estimation: Case of TdP's IAP (in thousands of US$)

	Unit cost	Q	Year 0 Cost
Hardware			
Router	26.0	1	35
Server	35.0	2	70
Workstation	3.0	10	30
Switch-hub	5.0	1	5
Modems*	0.2	4	1
Terminal servers	6.0	1	6
Printers	2.0	2	4
Scanner	2.0	1	2
CD-ROM	0.4	5	2
Digital camera	1.5	2	3
Tape backups	1.5	2	3
Instruments	260.0	0	0
Spare parts	25.0	0	0
Other	25.0	0	0
Software	50.0		50
Solaris 2.4			0
CISCO Works			0
SunNet Manager			0
Graphics			0
Programs			0
Officemate			0
Training	30.0	1	30
Total			240

Note: Q is quantity.

NONCOMPETITIVE COSTS

An IAP of TdP needs to lease only Infovia and Unired services. It does not need to lease national or international long-distance dedicated circuits or commercial telephone lines. In our estimation the IAP would only need to rent one E1 connected to Infovia and one E1 connected to Unired during the first 2 years. In the third year, it would need to lease an additional E1 each for Infovia and Unired (Table 3.9).

COMPETITIVE COSTS

Competitive costs were also lower in this case because an IAP requires less infrastructure than an independent ISP.

- *Capital investment.* This is highly reduced for an IAP, because most of its infrastructure is shared with TdP.[17] Thus the total capital investment would be $241,000 (U.S.), which represents just 10% of the total investment for a rival ISP (Table 3.10).
- *Other costs.* By assumption, they remain the same as before (Table 3.11).

TABLE 3.11
Costs Estimation: Case of TdP's IAP (in thousands of US$)

Noncompetitive costs	Year 0	Year 1	Year 2	Year 3	Year 4	Year 5
Installation tariffs	2	0	2	0	0	0
Unired via dedicated circuit	1	0	1	0	0	0
Infovia via frame relay	1	0	1	0	0	0
Monthly rentals	0	183	183	366	366	366
Unired via dedicated circuit	0	97	97	194	194	194
Infovia via frame relay	0	86	86	172	172	172
Total noncompetitive costs	2	183	185	366	366	366

Competitive costs	Year 0	Year 1	Year 2	Year 3	Year 4	Year 5
Investment	261	0	0	0	0	0
Capital equipment	241	0	0	0	0	0
Installation	20	0	0	0	0	0
Operating costs	0	378	492	546	558	630
Wages	0	204	318	372	384	456
Rent of building	0	72	72	72	72	72
Energy	0	30	30	30	30	30
Materials	0	72	72	72	72	72
Bad debts (1% of revenues)	0	7	20	32	45	57
Income tax (30% of gross profits)	0	0	83	299	569	821
Labor sharing (10% of operating profit)	0	0	28	90	190	274
Other expenses	0	180	180	180	180	180
Total competitive costs	261	565	803	1,147	1,541	1,961
Total cost	263	748	989	1,513	1,907	2,327
Ratio: noncompetitive/total	1%	24%	19%	24%	19%	16%

Column 2 of Table 3.12 shows the incremental cost for an IAP of TdP estimated in this alternative scenario. The incremental cost of the noncompetitive input was $0.091 (U.S.) per megabit, a little higher than TdP's cost data, but definitely still 80% below our estimate for an independent ISP.

---

17. For instance, TdP would keep IAP's routers in its exchanges. This advantage is only enjoyed by TdP's IAPs but not by the competitors.

TABLE 3.12
Incremental Cost Estimates: Case of TdP's IAP (US$ per Megabit)

	Bottom-up model	TdP's IAP	Comparison %
	(1)	(2)	(3)=(2)/(1)
(a) Noncompetitive cost	0.454	0.091	20
(b) Competitive cost	0.566	0.385	68
Total (a) + (b)	1.020	0.475	47

Notes:
(1) Results from the bottom-up cost model. Source: Column 1, Table 3.7.
(2) Results from the bottom-up cost model. Source: Table 3.9, 3.10, and 3.11.

These results suggested that TdP had been unlawfully favoring its affiliate firms in the access service to the Internet by charging them lower rates for noncompetitive inputs than their rivals. In other words, TdP would have been illegally increasing costs of its competitors in the market of Internet access to undermine competition. The potential harmful effect of this practice is to undermine competition and so extend TdP's dominant position to the competitive market for access to the Internet. In the same way, the results provide some evidence that TdP has been implementing cross-subsidies in favor of final services of access to the Internet by transferring resources out of regulated business such as dedicated circuits and basic telephone.[18]

CONCLUDING REMARKS

The presence of a vertically integrated telecommunications firm that is at the same time a dominant firm in most of its markets and faces no legal barriers to entry into any telecommunications service raises not only theoretical but also practical and regulatory concerns like those presented in this chapter. TdP, the Peruvian incumbent provider that entered the Internet market in 1997, sells at the same time noncompetitive or essential services to its downstream competitors in the Internet access market. Allegations of TdP's anticompetitive behavior came from established ISPs. One of these allegations held that TdP was practicing price discrimination in essential services to its rivals in the downstream market. Of course, TdP always denied such allegations. When OSIPTEL's intervention was requested, the agency decided to undertake several areas of investigation at the same time. One of these avenues was to analyze whether signs of price discrimination could be shown by an imputation test.

---

18. If a subsidy exists, we consider that our estimates could be underestimating the true magnitude of it because the assumptions made in the calculations have been very conservative regarding the costs and rather relaxed regarding income.

The imputation test presented in this chapter relied on public information on tariffs and on an economic engineering model for an efficient firm providing access service to the Internet. From the results found in this investigation, it was concluded that TdP engaged in discriminatory practices against competitors in the Internet market. Evidence of discriminatory treatment by TdP was found in the provision of noncompetitive inputs, adversely affecting competitive conditions in the Internet access market. An empirical estimation indicated that for noncompetitive inputs the differential between the price that TdP was imputing to itself and the one it charged to its competitors was 80%. Furthermore, when our noncompetitive cost estimate was imputed to TdP cost figures, it resulted in a revenue shortfall for TdP's Internet services. To the extent that price differences in noncompetitive inputs may be explained by the presence of economies of scope, TdP would have to show that such economies cannot be achieved in the provision of noncompetitive inputs to its competitors.

A definitive assessment about the presence of discriminatory pricing, cross-subsidies, or economies of scope on the part of TdP could not be performed because there was incomplete disclosure of information about TdP's costs and network topology, among other things. OSIPTEL requested such information, but TdP refused to provide it.

This analysis was very simple and tried to capture the most salient features of Peruvian Internet service provision in 1996 and 1997. However, the analysis did contain some shortcomings. Certain costs incurred by dial-up subscribers have not been incorporated into the price discrimination estimation. For instance, when the three-digit number to access the Internet through Infovia is dialed, the user pays the local measured telephone tariff irrespective of his or her actual location in the country. In contrast, the average subscriber of one of the independent ISPs will end up paying more for the Internet access because a subscriber in a given province in which the independent ISP has no presence will pay for a long-distance call instead of a local call.[19] To avoid such discrimination, subscribers of a rival ISP should have the same right of access to a three-digit number and pay for it as a local call. Including the local/long-distance price difference in the analysis would have increased the degree of total discrimination found with the simple model. Similarly, in the case of dedicated subscribers, some costs have not been taken into account in our estimates. Thus, a dedicated subscriber of an independent ISP or TdP has to contract for the dedicated circuit service in addition, of course, to the Internet access service. However, the former is charged a two-end port monthly rental for the dedicated circuit service, but if the same subscriber were to get Internet access as a TdP client, he or she would pay only a one-end port monthly

---

19. Assuming that only one node of Infovia exists, and is located in Lima, all the user calls from the provinces would be using the national long-distance network and therefore these calls would be long distance.

rental for the dedicated circuit. To include this feature in the estimation would also have amplified the degree of discrimination found in this chapter.

ACKNOWLEDGMENTS

This chapter is a simplified version of the analysis of anticompetitive behavior in the Internet market undertaken by the author for OSIPTEL during the first half of 1997. An extended version of the analysis is contained in the report that OSIP-TEL's Regulatory Bureau prepared in September 1997 for the commission in charge of resolving the allegations of anticompetitive practices in the Internet market: *Controversia entre Red Científica Peruana (RCP) y Telefónica del Perú (TdP) sobre Competencia Desleal e Incumplimiento del Contrato de Concesión.*

I am grateful to Padmanabhan Srinagesh for his comments and his advice to undertake this empirical investigation, and to Daniel Shimabukuro, who was in charge of developing the engineering model for an ISP.

The views expressed in this chapter are those of the author and do not necessarily represent the opinions of OSIPTEL.

REFERENCES

Larson, A. C., & Parsons, S. G. (1993, Fall). Telecommunications regulation, imputation policies and competition. *Hastings Communications and Entertainment Law Journal* 16(1) (pp. 1–50).

Mitchell, B., & Vogelsang, I. (1998). Markup pricing for interconnection: A conceptual framework. In D. Gabel & D. F. Weiman (Eds.), *Opening networks to competition: The regulation and pricing of access* (pp. 31–47). Boston: Kluwer.

OECD. (1996). *Information infrastructure convergence and pricing: The Internet.* Paris: Committee for Information, Computer and Communication Policy.

Vogelsang, I., & Mitchell, B. (1997). *Telecommunications competition: The last ten miles.* Cambridge, MA: MIT Press/AEI Press.

Srinagesh, P. (1995). Internet cost structure and interconnection agreements. In G. Brock (Ed.), *Toward a competitive telecommunication industry* (pp. 251–274). Mahwah, NJ: Lawrence Erlbaum Associates.

Trademarks and Domain Names: Property Rights and Institutional Evolution in Cyberspace

Milton Mueller

Syracuse University

Researchers gathered facts about 121 known cases of trademark-based challenges to domain name registrations. The cases were categorized according to the type of conflict and the kind of settlement or decision that resulted. The data show that a large majority of the cases (88%) would not qualify as trademark infringement under traditional standards of case law. Only about 12% of the cases exhibited the kind of consumer confusion, intent to pass off, or dilution that would normally be considered a trademark violation. In all of the cases of passing off, trademark owners won decisive victories in court. The largest number of domain name trademark cases (49%) arose from conflicts over the use of common names (such as *prince* or *columbia*) that legally may be used concurrently by multiple organizations or businesses. In many of these cases, courts have allowed trademark owners to take away names from other Internet users. The bias toward trademark owners has been exacerbated by dispute resolution procedures used by domain name registries that privilege trademark rights over all other claims to the right to use a name. The chapter concludes that trademark interests are expanding the scope of their property rights in cyberspace at the expense of smaller Internet users. The chapter concludes by proposing changes in law and registry policies that would rectify these injustices.

INTRODUCTION

The Internet is currently in the throes of a global redefinition of its institutional framework. One of the major points of controversy in this process is the conflict between domain names and trademarks. This chapter is an empirical study of the domain name and trademark interaction. It examines the conflicts that have led to trademark-based challenges to domain name registrations, and the settlements or decisions that have resulted to show how a new system of property rights is being forged. I employ a hybrid of legal and quantitative methods. The study collected information about 121 cases of conflict between domain name registrants and trademark owners. These cases were classified into four distinct types based on the way the challenged domain name was being used. The study also categorized the way the cases were settled.

REGULATING DOMAIN NAMES TO PROTECT TRADEMARKS

A *domain name* is a hierarchically structured character string that serves as an Internet address. The "real" Internet Protocol (IP) address is a string of numbers that is difficult to use and remember (e.g., 128.82.75.52). Internet users rely on domain names, which take the form of memorable and sometimes catchy words, to stand in their place. The uniqueness requirement creates an exclusivity that has important economic consequences—no two users can use exactly the same character string as a domain name.

Domain names are interesting—and controversial—because they are meaningful. The semantic dimension allows domain names to act not only as unique addresses, but also as slogans, billboards, or brand names.[1] The business identity of many Internet firms, such as Internet guide Yahoo! or book retailer Amazon.com, is solidly linked to their domain names. Even non-Internet-based businesses want domain strings that match, or are easily derived from, their corporate names.

The character strings used for domain names sometimes correspond to the character strings of registered trademarks. Often this occurs accidentally. Juno Online Services registered juno.com but the Juno Lighting Company holds a federal trademark on the brand name Juno. Sometimes the correspondence is a deliberate result of name speculators.[2] It is fairly inexpensive to reserve a name ($35–$50 per year is the norm), and many character strings may be perceived as more valuable than the going price. Thus, some "cyber-squatters" attempted to make a business out of the practice of reserving desirable names and attempting to resell them to others, sometimes for thousands of dollars. Sometimes the correspondence between a domain name and a trademark is part of a pattern of deliberate passing off, which is intended to confuse or deceive customers.[3] There are also numerous gray areas. To the Internet, one minor difference in a character creates a completely dif-

---

1. Burk (1995) discusses some of the legal dimensions of identifiers and the similarities and differences between domain names and other identifiers, such as telephone numbers. See also Brunel (1996) and Dueker (1996).
2. See Quittner (1994).
3. See INTA (1997).

ferent address but to users the difference may be hard to discern. Thus, some small businesses register domain names that are the same as a famous name but with only one minor character difference (e.g., amazom.com or micros0ft.com).

All of these conflicts over domain name rights have resulted in litigation, or threats of litigation, under trademark law. Both the users who registered the name and the domain name registry itself have been targets of such litigation. Internet registries and policymakers have responded to trademark challenges by making or proposing numerous restrictions on the way Internet domain names are distributed or used.

Network Solutions, Inc. (NSI), for example, is the exclusive government contractor selected by the National Science Foundation to register names under the top-level domains (TLDs) *.com*, *.org*, and *.net*. NSI allowed trademark holders to trigger a suspension of a user's domain name without any court proceeding whenever the domain name character string corresponds to their registered trademark.[4] Suspension can occur regardless of whether any use is being made of the name that would correspond to the legal definition of infringement. A January 30, 1998, draft proposal of the U.S. Department of Commerce proposed a similar policy.[5] An even more extreme policy was proposed by the so-called gTLD-MoU group based in Geneva.[6] Their proposal for domain name reform attempted to give famous name holders a right to preempt all character strings similar to or corresponding to their name in all TLDs. The gTLD-MoU also proposed that all domain name distributions be subjected to an extensive review and challenge process administered by the World Intellectual Property Organization (WIPO). If that policy had been implemented, trademark owners could mount challenges to domain name registrations based entirely on the character string, regardless of how it was used. Although the gTLD-MoU's attempt

---

4. NSI dispute resolution policy has now gone through five iterations. The earliest versions allowed NSI to suspend a domain name registration 30 days after presentation of a trademark registration by a challenger. See http://www.patents.com/nsi.sht for a discussion of various versions of the policy and their legal and economic consequences.

5. "If an objection to registration is raised within 30 days after registration of the domain name, a brief period of suspension during the pendency of the dispute will be provided by the registries." Appendix 2—Minimum Dispute Resolution and Other Procedures Related to Trademarks, National Telecommunications and Information Administration, 15 CFR Chapter XXIII, Improvement of Technical Management of Internet Names and Addresses, Proposed Rule, *Federal Register*, February 20, 1998 (Vol. 63, No. 34), pp. 8825–8833.

6. The so-called gTLD-MoU group was the product of a political alliance between the Internet Assigned Numbers Authority (IANA), the Internet Society (ISOC), the International Telecommunication Union (ITU), and the World Intellectual Property Organization (WIPO). In addition to its elaborate domain name challenge procedures, the gTLD-MoU proposed creation of seven new generic TLDs as an alternative to the exclusive government contract held by NSI. over *.com*, *.net*, and *.org*. See http://www.gtld-mou.org/. For an analysis of the historical background of its emergence, see Mueller (1998).

to assert authority over domain names failed, its attempt to regulate name registration to enhance trademark protection survives in a proceeding underway at the WIPO. Encouraged by a U.S. government white paper asking it to study the problem,[7] WIPO is holding an international proceeding that is expected to propose extensive regulations on the distribution of domain names.[8]

Politically, trademark owners have been instrumental in stifling widespread demands by Internet service providers (ISPs) and users for an expansion of the supply of TLDs[9] Many trademark holders opposed the gTLD-MoU because of its creation of seven new TLDs. Among corporate trademark holders, the idea that the distribution of domain names should be subjected to some kind of regulatory process to protect trademark rights has become widely accepted.

Many legal analysts have been critical of the linkage between domain names and trademarks. DeGidio (1997) and Hilleary (1998) criticized the vagueness and potential for abuse of the new federal antidilution statute as it has been applied to domain name cases. Oppedahl (1997) argued that the domain name challenge procedure of NSI gives trademark holders more sweeping rights than they have under trademark law. Nathenson (1997) showed that the concept of dilution has been exploited by trademark holders in the domain name arena to engage in "reverse domain name hijacking."

Missing from both sides of this debate is a sense of the quantitative dimensions of the problem. The legal literature about domain names has already recognized that there are distinct types of domain name conflict and that the applicable law may hinge on what type it is.[10] We submit that it is important and useful to know the statistical frequency with which the different types occur. The quantitative information affects our identification of the problem, our assessment of its significance, and our conception of what might be an appropriate policy response.

Suppose, for example, that there were a half a million known instances of domain name conflict and that 80% of them could be readily classified as cases of

---

7. *White Paper on Management of Internet Names and Addresses,* U.S. Department Of Commerce, National Telecommunications and Information Administration, Docket No: 980212036-8146-02 (June 5, 1998).

8. See WIPO RFC-2, "Request for Comments on Issues Addressed in the WIPO Internet Domain Name Process," September 14, 1998. Available http://wipo2.wipo.int/process/eng/processhome.html

9. See, for example, the June 8, 1998 news release of the International Trademark Association, which states "The creation of additional gTLDs is an issue the Association has fought since the debate on the Internet began several years ago." Sally Abel, Vice Chair, INTA Issues and Policy Committee, June 8, 1998, PR Newswire.

10. Nathenson (1997) for example, developed a three-category classification scheme (squatters, twins, and parasites) that has some similarities to the one proposed here. Oppedahl (1997) laid out a list of factors regarding the type of name and its use that affect legal remedies.

clear and intentional infringement. On the basis of such quantitative evidence, it would be reasonable to conclude that trademark infringement using domain names was a major problem on the Internet. But suppose that the number of cases of conflict is much smaller and that the vast majority of domain name–trademark conflicts arise from name speculation. Such an outcome would lead to an entirely different conception of the problem and suggest different policy responses. Perhaps the profitability of name speculation depends on current artificial limits on the number of TLDs, and public policy should be focused on the need to expand the name space. Or perhaps registries should impose a "use it or lose it" policy and require immediate payment to prevent speculators from hoarding names. Alternatively, suppose the data suggest that most of the conflicts arise from innocent attempts to use the same word by businesses with no intent to deceive or infringe. If this type of conflict was most common, the notion that domain name conflicts are primarily a trademark problem is powerfully undermined.

METHOD

The following method was developed to identify the statistical parameters of the problem:

- The researchers identified and collected as many cases of domain name conflict involving trademark claims as possible. Cases were identified on an entirely unselective basis by a research assistant who was not responsible for their classification. A case was then included in or excluded from the database whenever sufficient information about it could be gathered to classify it properly.
- The researchers defined an exhaustive, mutually exclusive set of categories that describes (a) the type of conflict, and (b) the type of settlement (see following for elaboration of the categories).
- Each case was classified with respect to the categories defined earlier.
- Finally, the researchers counted the frequencies of the various categories and developed cross-tabulations between the categorization schemes (conflict type and settlement type).

The unit of analysis in this study is the domain name itself. Each domain name registration that was challenged on trademark grounds constituted a single entry in the database. Thus, if a single lawsuit encompassed two or three names, each challenged name counted as a single case in our collection of data. It should also be noted that the field of study was inherently limited to cases that received some kind of public notice, meaning either published reports in printed and online media or records of court proceedings.

In the search for cases the method had to determine what should and should not be counted as an instance of domain name–trademark conflict. Inclusion was based on the case's ability to meet three criteria: (a) there must be a challenge, (b) the challenge must be focused on the domain name itself, and (c) the challenge must be based on trademark rights. Challenges could take the form of a letter, a telephone conversation, or any other form of notification that was reliably reported; it did not just include lawsuits.

CATEGORIZATION OF CASES

Once cases were included in the database they were categorized according to two parameters. The first parameter classified the type of conflict. We identified four distinct categories of conflict: infringement, speculation, character string conflicts, and parody and preemption.

The following sections elaborate on the meaning of these categories and construct an argument that they are reasonably clear, mutually exclusive, and exhaustive as a categorization scheme.

Infringement

Infringement refers to domain name conflicts in which the original registrant intentionally attempts to trade off the resemblance between the domain name and another company's trademark. Also included under this category are so-called dilution cases in which the value of a mark would be tarnished or devalued by its association with the site of the domain name user. In short, these are uses of a mark that would be illegal under existing trademark concepts, regardless of whether they occurred as Internet domain names or in any other context. Such cases can be readily identified by the application of a standard checklist for identifying infringement that has emerged from case law.

The standard factors pertinent to a finding of trademark infringement include: (a) the strength of the trademark, (b) the similarity between the plaintiff's and defendant's marks, (c) the competitive proximity of the parties' products, (d) the alleged infringer's intent to confuse the public, (e) evidence of any actual confusion, and (f) the degree of care reasonably expected of the plaintiff's potential customers.[11] The new federal antidilution law[12] also creates a cause of action for the owner of a famous mark where there is no competition between the parties and no confusion as to the source, but when the use of the name would diminish the distinctiveness of a mark. The name in question has to be nationally "famous" and there still must be "usage in commerce."

---

11. *Anheuser-Busch, Inc. v. Balducci Publications,* 28 F.3d 769, 774 (8th Cir. 1994); see also *Centaur Communications,* 830 F.2d at 1225; *Duluth News-Tribune,* 84 F.3d at 1096.

12. Federal Trademark Dilution Act of 1995, 15 USC 1125 (c).

Speculation

Name speculation occurs when a domain name bearing a resemblance to a registered trademark is registered as an act of speculative arbitrage. In this type of case the name is not used by the speculator, but is held in reserve in the hope that the trademark owner will eventually buy it from the person who registered it. Name speculation is treated as a distinct category of domain name conflict that should not automatically be grouped with cases of infringement. Because speculators generally do not use the name and because of their obvious intent to resell the name, name speculation cases are generally easy to identify and categorize.

Name speculation is often equated with trademark infringement, both in the press and in the courts. This equation stretches trademark protection beyond its traditional limits, however. The hallmark of name speculators is that they do not use the names; that is, the domain names are not part of a commercial product or service offering that might confuse or deceive customers or undermine the distinctiveness of a mark. Thus, the most important criteria of trademark infringement— use in commerce, likelihood of confusion, and dilution—are completely absent from speculation cases. No one can be confused by a blank screen. The distinctiveness of a mark cannot be devalued by a nonexistent website. Name speculators never make any pretense of being the company whose name they control. They are simply trafficking in the name resources themselves.

Admittedly, the contention that name speculation is not infringement per se runs counter to the general thrust of recent legal decisions. As I show in the Results section, court opinions regarding name speculation in the United States, United Kingdom, and New Zealand have uniformly declared it to be an illegal form of trademark dilution.[13] This does not affect the validity of our categorization scheme. Name speculation constitutes a distinct type of trademark–domain name conflict, one that is clearly distinguishable from the traditional form of infringement contemplated by pre-Internet trademark law. Indeed, the classification of name speculation as trademark infringement constitutes a novel and expansive interpretation of trademark protection, and all legal decisions regarding name speculation have been forced to recognize this. The court in *Panavision v. Toeppen,* for example, stated, "Registration of a trademark as a domain name, without more, is not a commercial use of the trademark, and therefore is not within the prohibitions of the [Dilution] Act."[14] The U.K. court was forced to acknowledge that name speculation itself was not infringing, but ruled it illegal because it created an "obvious threat" of some form of infringement in the future.[15] In short, even the

---

13. *Panavision International v. Toeppen,* 46 U.S.P.Q.2d 1511, 1998 WL 178553 (9th Cir. April 17, 1998); *Intermatic Inc. v. Toeppen,* 947 F. Supp. 1227 (N.D.Ill. 1996); *Cadbury Confectionery & Ors v. The Domain Name Company Ltd & Ors* (New Zealand). In the New Zealand case consent orders were given but no written judgment was issued.

courts that ruled against speculators found it impossible to classify their actions as a form of simple, classical infringement; they had to recognize its unique status and develop novel (often spurious) rationales to declare it illegal. Thus, it is important to retain a separate classification for such cases.

Character String Conflicts

Character string conflicts occur when there is more than one legitimate, non-speculative user of a given character string as a domain name. In string conflicts, there is no apparent intention to trade off a trademarked name and little or no potential for confusion between the products of the conflicting companies. Both parties either have a trademark of their own or a valid reason to use a particular character string (e.g., the string corresponds to a nontrademarked company name, product name, or personal name). String conflicts often involve company names based on generic terms (e.g., *disc, prince,* or *perfection*) that may be trademarked in many different industrial categories or jurisdictions. String conflicts occur because domain names strip away many of the unique characteristics of product names and logos and because the global reach of the Internet transcends many of the jurisdictional and geographic limitations placed on traditional trademarks. For example, Oppedahl (1997) noted that the word *glad,* which formed the basis of a domain name dispute, is a trademarked name of more than 200 businesses in the United States, plus numerous foreign trademarks, in industries ranging from cosmetics to electrical apparatus to detergents. It is evident that a trademark holder may claim trademark infringement or dilution to win a string conflict. Such a claim may be sincere, but is more likely to be a way of taking a desired domain name from someone who registered it first. Generally, it was not difficult to distinguish real infringement cases from string conflicts because of the standards developed by traditional case law. Whenever there was any doubt about whether to classify a case as infringement or a character string conflict, we tried to err on the side of infringement.

---

14. Panavision International, L.P., a Delaware Limited Partnership, Plaintiff, v. DENNIS TOEPPEN, an individual, Network Solutions, Inc., a District of Columbia Corporation, and DOES 1-50, Defendants. Case No. CV 96-3284 DDP (JRx) November 1, 1996. Nathenson (1997) wrote that name speculation cases are "more legally problematic than the courts have made them appear" and DeGidio (1997) criticized the *Intermatic* court's classification of name speculation as "use in commerce."

15. Deputy Judge Jonathan Sumption QC said in the U.K. case that mere registration of a name was not, in itself, passing off or infringement of a trademark, but that injunctions should be granted against name speculators "to restrain the threat of passing-off" (BBC News, November 28, 1997).

Parody, Preemption, and Other

This category turned out to be a small minority of cases, but as a group most of them bore a distinct family resemblance. These cases generally involve uses of domain names for acts of parody, preemption, or expression. In this type of conflict, the domain name deliberately invokes or resembles a company name or trademark to make a political or satirical point. Someone registered the name british-telecom.com to post information critical of that company, for example. Or the domain name itself may make a comment on the organization (scientology-kills.net). Perhaps the most famous parody case is the peta.org conflict, in which the acronym for People for the Ethical Treatment of Animals (PETA) was registered by a critic who set up a satirical site for a fictitious organization called People Eating Tasty Animals. In a slightly different vein, domain names can be registered to prevent someone else from using them. In some cases, these preemptive registrations are not used at all; in other cases the domain name will be used to redirect traffic to a completely unrelated site. Thus, for example, the domain name ringlingbrothers.com is registered to PETA, which at first used it to post information critical of circuses' treatment of animals, and later to redirect traffic to an environmentalist site.

These parody and preemption uses of domain names were not classified as infringement unless the domain name user pretended to be the organization that is being criticized, spoofed, or preempted. If deception was intended or confusion was likely, we classified the case as infringement.[16] Also, most of the cases of parody and expression are noncommercial uses and therefore would not be liable to infringement claims under traditional law. The cases in this category do not qualify as a string conflict either, because the resemblance between the domain name and the affected company's name is not innocent or unintentional. Of course, to be included in the study at all, domain name conflicts in this category had to be challenged on trademark grounds. There are some known cases of parody use of famous names that have not (yet) been challenged.[17]

CATEGORIZATION OF OUTCOMES

In addition to classifying the type of case, we classified the results. Results were categorized according to the following schema. We were interested primarily in

---

16. Thus, the plannedparenthood.org case was classified as infringement rather than parody, because the website run by the antiabortion activist claimed to be the "home page" of the Planned Parenthood organization.

17. For example, the domain micr0soft.com (the first "o" is the numeral 0) is registered to the "Microsoft Sucks" organization, and its home page is emblazoned with the slogan, "What do you want to hack today?" As far as we know, Microsoft has not mounted a legal challenge to it.

whether the original registrant or the challenger retained the right to use the name after the challenge. This determination could occur in one of four ways (see Table 4.1).

TABLE 4.1
Categorization of Results

	Court decision	Registry action	Settled during litigation	Settled without litigation
Original registrant retains name	A		D	F
Challenger wins name	B	C	E	G
Pending/unknown		H		

- First, a formal court decision could uphold the original registrant (A) or the challenger (B).
- Second, the registry could suspend the name, thereby destroying the usage right of the original registrant[18] (C).
- Third, disputants could reach a private settlement after litigation was initiated (D, E).
- Fourth, disputants could reach a private settlement without litigation (F, G).

The classification was based on the final disposition. Thus, in many cases, NSI may have suspended or threatened to suspend a name, but this action was ultimately superseded by a formal court decision. In that case the result would fall into Category A or B rather than C. Similarly, if a case was settled after a suspension, it would be classified as D, E, F, or G rather than C. There was also a residual category for outcomes that were pending or unknown (H).

RESULTS

According to NSI, of the nearly 2 million domain names it had registered as of May 1998, only 3,903 had been the subject of a complaint. Of those, only 2,070 invoked NSI's trademark dispute resolution policy, and only 1,523 were placed on hold as a result. This means that a very small fraction of all domain name registrations (0.00104) created a trademark conflict.[19]

The results of the classification exercise were surprising. We entered into the study with the expectation that name speculation would be the chief cause of

---

18. Suspension was classified as "challenger wins" because it prevents the original registrant from using the name at the behest of the challenger. Technically, however, NSI does not transfer the usage right from one party to another without a court order, so in cases of suspension the challenger is not actually awarded the name. However, suspension can coerce a registrant to surrender the name to a challenger.

19. E-mail from David Graves, NSI, to Milton Mueller, May 4, 1998.

trademark conflicts. We were wrong. String conflicts, wherein two companies with a roughly equal claim to a name contest the right to a domain registration, was by far the most common type of case, making up 49% of the cases. Conflicts arising from name speculation came in a distant second, at 35% of the cases. Cases of infringement constituted only 12% of the sample. Parody and preemption cases made up only 4% of the cases (see Table 4.2). If these proportions are projected into the NSI figures about the total number of trademark disputes it has handled, one could estimate that of the 2 million or so domain names registered, there are probably only about 257 cases (2,070 x 0.124) of infringement. This amounts to only 0.0128% of all domains. Given this miniscule percentage, it is truly remarkable that trademark interests have been able to portray trademark infringement as a pervasive problem that should be one of the primary drivers of domain name distribution policies.

TABLE 4.2
Conflict Types

Type of conflict	Number of cases (N=121)	Percentage of total sample
String	59	48.8
Speculation	42	34.7
Infringement	15	12.4
Parody/preemption	05	4.1

Note the dominance of string conflicts in the results. Disputes over character strings involve second-level domain names such as *clue, disc, pike, newton, glad, compassion,* and apparently innocuous acronyms such as *dsf* or *dci.* That such cases constituted the largest single category of domain name–trademark disputes has important policy implications. String conflicts are the type to which the application of trademark law is least appropriate. Trademark law permits concurrent use of the same or similar marks in different industries or different geographic regions. Traditional trademark protection accords generic terms much less protection than fanciful, invented names. Clearly, trademark claims are being inserted into conflicts over Internet resources where they do not belong. Indeed, so-called dispute resolution policies such as NSI's actually create legal conflict by encouraging trademark owners to claim names they did not get first. The possibility of trademark litigation, or name suspension by a registry, allows parties with greater legal resources to take desirable names away from legitimate prior registrants. These conclusions are reinforced when the typology of cases is connected to the outcome categories.

When domain names are used in a way that conforms to traditional notions of infringement, the data show that traditional legal remedies provide strong and relatively quick protection. We classified 15 (12.4%) of the 121 cases as infringement. In 14 of the 15 cases (93%), the trademark holder who challenged the domain name succeeded in winning it back or preventing the infringing party from using it. In none of these cases was it necessary to utilize the registry to put the name on hold. Ten of the cases were decided in court. Two were settled in favor of the challenger immediately after a lawsuit was filed, but before any court decision was necessary. Two were settled in the trademark holder's favor prior to any litigation (see Table 4.3).The one case in which infringement was present but the original party retained the domain name did not represent a defeat or diminution of the rights of the trademark holder. A settlement between the trademark holder (Digital's Alta Vista Internet search engine) and the domain name holder (a software company called Alta Vista, holder of the altavista.com domain) prevented the defendant from pretending that its website was an affiliate of Digital's Alta Vista search engine (altavista.digital.com). It required that a disclaimer be added to the defendant's home page. Digital's settlement thus recognized that there was no inherent infringement in the Alta Vista company's use of the domain name itself, even though user confusion existed and the defendant company had been exploiting it. The settlement merely prevented infringing behavior.

TABLE 4.3
Infringement Cases: Outcome Categories

	Court decision	Registry action	Settled during litigation	Settled without litigation
Original registrant retains name	0		1	0
Challenger wins name	10	0	2	2
Pending/unknown		0		

In short, when real trademark infringers utilize domain names as part of their tactics, empirical evidence suggests that litigation provides a nearly 100% effective remedy. The evidence also shows that these disputes have been resolved rather quickly, except in one case.

The exception is important. A clear infringer has managed (for the time being) to retain a domain name despite an adverse court decision. A Spanish-language search engine posted at the domain ozu.es sued an imitator using the domain ozu.com. The infringing site, ozu.com, ran the exact same type of business as ozu.es and even posted an exact reproduction of the logo of the other site. A lawsuit was filed and a Spanish court resolved the case in favor of ozu.es.[20] But as of mid-May, 1998, the ozu.com domain remains in the hands of the infringing party. For reasons known only to itself, NSI refused to comment on whether it is aware

20. JUZGADO DE PRIMERA INSTANCIA N° 13 BILBAO Lerkundi, 20 Procedimiento: Medidas Cautelares De: Advernet, S.L Contra: OZUCOM, S.L y Don E.

of the case or whether it has received an order from the Spanish court to transfer the domain. British court decisions ordering name transfers in NSI's *.com* and *.org* TLDs, however, appear to have been implemented quickly by NSI.

This problem does not appear to be a major one. According to NSI, of the 3,903 complaints it has received of domain names, only 45 involved trademarks from foreign (non-U.S.) jurisdictions. Rather than making domain name registries miniature trademark courts, problems such as this could be addressed by international agreements among registries to respect the court decisions of other jurisdictions. Another point to keep in mind is that cross-jurisdictional intellectual property enforcement is imperfect outside of cyberspace as well. We uncovered four major instances of name speculation: two in the United States (Dennis Toeppen and Jim Salmon), one in the United Kingdom (R. Conway & J. Nicholson), and one in New Zealand (David Ward). There were other smaller cases in the United States, Hong Kong, and Australia.

Name speculation fared poorly among courts and registries (see Table 4.4). Of the 36 cases for which outcomes were known, 33 of the contested names (92%) ended up in the hands of challengers. Only three of the cases of speculatively registered names remain in the hands of name speculators; these are also three cases that have not yet been challenged in court. We therefore classified them as settled because we concluded that the businesses affected do not care enough about the problem to pursue litigation. Eighteen (50%) of the names were taken away by formal court decisions. Not a single court decision in the sample upheld the right of name speculators to profit from their prior registration of a company name. Another eight cases (19% of the total) nullified the domain names through a suspension of the name by the registry. Seven cases were settled in favor of the challenger after a lawsuit was filed; that is, the speculator caved in without a legal battle. (In the remainder of the cases the outcome is unknown or unresolved.)

TABLE 4.4
Name Speculation: Outcome Categories

	Court decision	Registry action	Settled during litigation	Settled without litigation
Original registrant retains name	0		0	3
Challenger wins name	18	8	7	0
Pending/unknown		0		

An earlier section noted that it is difficult to classify name speculation as trademark infringement. Despite this fact, trademark owners who challenge name speculators, either in court or in name registries, have won every single case. The courts have not merely enjoined the use or sale of the names by the speculators; they have also transferred ownership of the names to the challengers. This has happened in the United States even though the federal dilution statute, which specifies detailed remedies for dilution, does not authorize the transfer of a domain name as a remedy. The fate of name speculation provides one of the clearest examples of the way in which trademark rights in cyberspace are being expanded beyond their normal meaning in law.

TABLE 4.5
String Cases: Outcome Categories

	Court decision	Registry action	Settled during litigation	Settled without litigation
Original registrant retains name	15		10	7
Challenger wins name	7	5	7	2
Pending/unknown		5		

TABLE 4.6
Parody Cases: Outcome Categories

	Court decision	Registry action	Settled during litigation	Settled without litigation
Original registrant wins	0		0	0
Challenger wins	0	1	2	1
Pending/unknown		1		

The domain name bt.org provides a poignant example of this process. Name speculators in Britain registered bt.org and several other names related to British Telecom. The British court did not merely enjoin the speculators from using the names, it also ordered that the bt.org name be given to British Telecom. This occurred despite the generic character of the acronym bt, despite the fact that British Telecom already had numerous domain names corresponding to its trademarks, and despite the fact that the .org TLD is generally used for the registrations of noncommercial organizations. The outcome pattern of the string cases (see Table 4.5) shows the confusion inherent in any attempt to apply trademark rights to parties with roughly equal claims to name resources. A majority (60%) of the known outcomes favor the original registrants (32–21). That becomes a better than two to one majority (15–7) in favor of the original registrant when string conflicts go all the way to a formal court decision. However, the majority of cases never get that far. There were five adverse decisions by domain name registries (primarily NSI). There were also seven adverse court decisions, some of them palpably unjust.[21] Those decisions, coupled with the tendency of some defendants to be intimidated into unfavorable settlements after a lawsuit is filed, all mean that an innocent domain name registrant's chance of retaining a domain name challenged by a trademark holder is only 10% better than flipping a coin. In the parody and preemption category (see Table 4.6), the number of cases is too small to make any robust conclusions possible, but a general pattern is discernable. Disturbingly, the original registrant retains the name in only one of the identified cases (scientology-kills.net), a case in which the outcome is pending. The remaining cases reveal a tendency to strengthen trademark protection at the expense of expression. Peta.org, for example, was placed on hold by NSI. The lawsuit mounted by Ringling Brothers and Barnum and Bailey against PETA regarding ringlingbrothers.com has led to a settlement in which PETA will surrender the name. The man who registered british-telecom.org to criticize the company was sued and relinquished the name.

CONCLUSIONS AND ANALYSIS

Real trademark infringement using domain names is a rare and not very significant problem: 0.0128%, or a total of no more than 257 cases in the generic TLDs administered by NSI. Nearly all infringement activity has been quickly stopped by lawsuits. Trademark protection is important, but it need not be one of the primary considerations in developing policies for domain name registries.

That the problem has been blown out of proportion is highly significant. It reflects the disproportionate political power of intellectual property holders. Trademark owners have been able to claim property rights in Internet domain names that go far beyond the rights they have under existing legislation and case law governing trademarks. They have been able to project into the "green field" of cyberspace a new conception of what they think their property rights in names ought to be. That conception, which extends trademark protection to string conflicts, name speculators, and even satirical and parody uses that under normal circumstances might invoke First Amendment protection, provides them with broader and stronger rights than they currently have in the physical world. Because the economic characteristics of Internet-related resources are new and unfamiliar, the courts have often (but not always) acquiesced in these extralegal claims.

More disturbing than the court decisions is the attitude of key Internet policymakers in the United States and Europe. They have shown a marked tendency to

---

21. See, for example, *Pike et al v. Network Solutions et al,* US Dist Court for North Dist of California (S.F.) 96-CV-4256, November 25, 1996. The court took pike.com away from Peter Pike, a real estate consultant in California, and awarded the name to Floyd Pike, an electrical contractor who repairs power lines in North Carolina. The trademarked logo of the electrical contractor used stylized letters that cannot be reproduced in a domain name. It is difficult to understand the legal rationale. How can a real estate consultant in California be confused with an electrical contractor in North Carolina?

use the centralized institutional mechanisms of Internet administration to preserve and advance extralegal conceptions of trademark rights. So-called dispute resolution procedures are essentially ways to exploit the bottleneck power of domain name registries to aid in expanding the scope of trademark rights and to socialize the costs of policing trademarks. To reduce their transaction costs, the trademark interests have opposed any attempt to create new TLDs. The entire market for Internet domain names, in other words, is being artificially constrained at the behest of trademark owners.

This expansion of trademark rights sets a dangerous precedent for the evolution of the Internet. It indicates that the central organizations used to allocate key Internet resources (domain names and numerical addresses) can be captured by business interests with a predominantly regulatory agenda. When Internet governance incorporates a regulatory agenda, the distribution of Internet domain names is no longer a matter of simply coordinating potentially conflicting claims. Rather, the distribution of these essential resources gradually becomes linked to control of conduct. Reacting to strong political pressures, the organizations governing the Internet are threatening to exploit the leverage they obtain from their control of domain names to extend public regulation into broader areas of Internet activity. The process is directly analogous to the early stages of radio broadcast regulation in the 1920s, when governmental control of the allocation of radio frequencies became the entering wedge for comprehensive controls on broadcast ownership, content, and conduct. In the immediate case, the distribution of domain names is being used to gain leverage over Internet users and suppliers, leverage that is then used to broaden the property rights of multinational corporate trademark holders. The domain name–trademark conflict thus has relevance that goes far beyond intellectual property rights. The decisions being made now about how to handle that conflict will have a decisive impact on the future of global computer networking. The domain name–trademark conflict is not a minor sideshow, but a watershed moment in the evolution of this global medium.

POLICY AND LEGAL PROPOSALS

Important revisions need to be made in the policies and laws regarding domain name–trademark conflicts. Current law and policy clearly recognize that domain names can be used to violate legitimate trademark rights. That recognition must be balanced with an explicit recognition that trademark claims can be used to abuse legitimate domain name usage. Law and policy must also recognize that there are important freedom of expression issues implicit in our handling of domain names. With these principles in mind, the following proposals are advanced.

First, and most important, name registries should be confined exclusively to the registration and transfer of domain names, and registries must be indemnified against legal liability for trademark violations by their customers. This is important, because if they are not so indemnified, they will be forced to adjudicate and police trademarks. If that happens, the rational course of action for domain name registries will be to implement conflict resolution policies that presumptively favor large trademark holders. Famous trademarks tend to be held by multinational corporations with the legal resources to create a great deal of trouble for registries. Smaller businesses and individuals cannot mount such a threat. If the third-party legal responsibility for trademark violations is thrust on registries, the methods they choose to resolve challenges will be designed to minimize their costs, and the easiest way to minimize costs is to let the big trademark interests win at the expense of all other Internet users.

Second, to protect the rights of domain name users, the federal dilution statute should be clarified to prevent attempts to resolve string conflicts via trademark litigation. The law must explicitly recognize that the character strings used as Internet identifiers do not and cannot reflect many of the distinguishing characteristics of real trademarks, such as geographic scope, jurisdictional differences, or typographical differences. String conflicts among legitimate claimants are therefore inevitable. In the resolution of these conflicts, trademark law does not provide appropriate criteria for the resolution of a conflict; it merely fosters needless legal conflict and unfair outcomes. Trademark or dilution concepts should not allow mark owners to assert a blanket right to control a character string regardless of how it is used.

Name speculation poses a more difficult problem. There is little doubt that certain forms of cyber-squatting have no redeeming social value. On the other hand, many innovative and successful Internet-based businesses are dependent on the ability to buy, sell, or arbitrage names. It may not be possible to prevent the former without also interfering with the latter. Consider the activities of Free View Listings, Ltd., a Canadian firm that registered hundreds of common surnames and resold them on a shared basis to individuals who wished to use their family name as an Internet address. This small company was sued by Avery-Dennison on trademark violation grounds because it registered the names avery.net and dennison.net, both ordinary surnames in English. No real infringement was proven. The Avery-Dennison company already had a well-known, well-established domain at avery.com. A benighted U.S. district court, caught up in the irrational frenzy against name speculators, classified Free View's registration and reselling of names as a form of cyber-squatting and ruled it illegal. Unless it is overturned on appeal, the decision will destroy a perfectly legitimate business. Trademark law was never intended to grant mark owners such sweeping, indiscriminate rights over the use of words. In name speculation cases as in string conflicts, any application of trademark rights to cyberspace must be balanced. Name speculators can abuse trademark owners, but trademark owners can and do use trademark claims to abuse legitimate forms of Internet trade.

Name speculation is really an economic issue and is best addressed by economic means. Its roots lie in the gap between the low price of reserving a name and the high value of certain names. The problem is exacerbated by the artificial limits on the number of generic TLDs and the privileged position of names under .com that result from this artificial scarcity. The creation of many additional TLDs would expand the supply of second-level domains, thereby reducing the profit and increasing the risk of name speculators. Such a course of action would also make it more difficult for trademark owners to control all uses of the character strings they are interested in. But in this regard, trademark owners may be seeking a level of control over words that the Internet, by its very nature, renders impossible.

ACKNOWLEDGMENTS

I wish to acknowledge my graduate research assistant, Ms. Heather Beach, and the helpful comments of Eugene Volokh, Sharon E. Gillett, Gervaise Davis, David Graves, and Jessica Litman. All responsibility for the content of this chapter rests with the author.

The complete list of cases, classifications, and results can be found at http://istweb.syr.edu/~mueller/list.html.

REFERENCES

Brunel, A. (1996). Trademark protection for Internet domain names. In Computer Law Association (Ed.), *The Internet and business: A lawyer's guide to emerging legal issues.* Fairfax, Va: The Computer Law Association. Available http://cla.org/RuhBook/chp3.htm.

Burk, D. L. (1995). Trademarks along the Infobahn: A first look at the emerging law of cybermarks. *Richmond Journal of Law & Technology, 1*(1). Available http://www.urich.edu/jolt/v1i1/burk.html.

DeGidio, A. J., Jr. (1997). *Internet domain names and the Federal Trademark Dilution Act: A law for the rich and famous.* Available http://www.lawoffices.net/tradedom/sempap.htm.

Dueker, K. S. (1996, Summer). Trademark law lost in cyberspace: Trademark protection for Internet addresses. *Harvard Journal of Law and Technology, 9,* 483.

Hilleary, S. (1998). *Cyber-dilution: Protecting domain names under the Federal Dilution Act of 1995.* Available http://www.gte.net/shawnah/cyberdln.htm.

International Trademark Association. (1997). *INTA white paper: Trademarks on the Internet.* New York.

Mueller, M. (1998). The battle over Internet domain names: Global or national TLDs? *Telecommunications Policy, 22*(2), 89–108.

Nathenson, I. S. (1997). Showdown at the domain name corral: Property rights and personal jurisdiction over squatters, poachers, and other parasites. *University of Pittsburgh Law Review, 58*. Available http://www.pitt.edu/~lawrev/58-4/articles/domain.htm.

Oppedahl, C. (1997). Remedies in domain name lawsuits: How is a domain name like a cow? *John Marshall Journal of Computer & Information Law, 15*, 437.

Quittner, J. (1994, July). Billions registered: Right now, there are no rules to keep you from owning a bitchin' corporate name as your own Internet address. *Wired, 2*(10).

II

REGULATION

An Economic Analysis of Telephone Number Portability

P. Srinagesh
Bridger M. Mitchell
Charles River Associates

We analyze several implications of local number portability (LNP) for competition and economic efficiency. In the first section we present a brief introduction to telephone numbers, the private and external benefits of number portability, and the costs of alternative architectural approaches to implementing number portability. We then describe cost recovery principles adopted by regulators in the United States, the United Kingdom, and Hong Kong. We next discuss the incentives of incumbent operators to engage in anticompetitive behavior and whether an incumbent's benefits from maintaining a stable directory infrastructure might diminish these incentives and internalize the external benefits of LNP. A brief concluding section summarizes our findings and identifies areas for future research.

TELEPHONE NUMBER PORTABILITY

The North American Numbering Plan

Every telephone (or more precisely, every main station) is assigned a unique address that is used for call setup and call routing and to identify the telephone for billing and other administrative purposes. The hierarchical structure of telephone numbers was developed, in part, to simplify call setup and call routing procedures. Until recently, the North American Numbering Plan (NANP) typically assigned each main station a number of the form N0/1X NNX XXXX. In this notation, N is any integer except 0 or 1, 0/1 is either the 0 or the 1 digit, and X is any integer. The first three digits (of the form N0/1X) are referred to as the *area code* and are shared by all numbers in a numbering plan area. The next three digits are called the *central office code*. A central office, which houses one or more switches, may be associated with several central office codes, but each central office code is associated with a unique central office. The last four digits are associated with individual lines or main stations. Today, NANP numbers are of the form NXX XXX XXXX, with the first three digits being the area code, the next three the central office code, and the last four the line number.

A virtue of this telephone numbering scheme is that it enables simplified call routing procedures, something that was particularly important for earlier generations of analog switches. Early switches could be programmed to recognize that a dialed number beginning with N (2–9) would consist of seven digits and could be routed locally, whereas a dialed number beginning with 0 or 1 would be followed by 10 digits and would need to be routed to the long-distance network. In general, switching logic could be simplified by structuring telephone numbers to reflect the hierarchical routing rules of the switched telephone network.[1] The structure of the telephone numbers has been efficient from another perspective as well: North American numbers are very densely used, and through judicious assignment, the move to 11-digit numbers to accommodate growing numbers of telephones has been deferred further into the future than would have been possible with less efficient schemes.

One consequence of efficient numbering systems developed by the integrated Bell system is the close link between numbers assigned to customers and the central offices (or switching systems) that serve the specific area where the customer is located. The concept of number portability—the ability of a customer to retain his or her telephone number while moving to a different location of telephone carrier—had only limited applicability. Traditionally, when customers moved outside the serving area of their wire center, they were assigned a new number with a central office code associated with his new serving wire center. Although customers who moved within a serving area were often able to keep their telephone numbers, geographic number portability was inherently limited to small areas around exchanges. Before divestiture, there was no need to "port" numbers across local exchange carriers because different local exchange companies did not share the same exchange areas. However, the lack of geographic portability was viewed by most observers as a minor inconvenience.

Benefits of Number Portability

With the advent of competing suppliers of local telephone service, the lack of LNP is viewed as a major impediment to effective competition. A customer of an in-

---

1. For a specific example, see the discussion of the Principal City Concept in *Engineering and Operations in the Bell System*, R.F. Rey, Technical Editor, AT&T Bell Laboratories, Murray Hill, New Jersey, 1984, p. 120.

cumbent local exchange carrier who decides to switch to a new entrant with a competing offering of local service will (with traditional technology) be required to obtain a new number based on a central office code associated with the entrant's central office. Most customers, especially those business customers who are disproportionately dependent on incoming calls, are reluctant to change their telephone numbers and incur the expense of notifying their customers and suppliers of the change. Lack of LNP across providers might therefore be a significant barrier to entry in the market for local service. LNP will remove this barrier to entry, and porting customers will not face the expenses of renumbering when they change providers. In addition to providing private benefits to porting customers, LNP has significant public good aspects and generates external benefits. These aspects of LNP complicate the economic analysis of LNP cost recovery.

Earlier analyses of LNP have identified both direct and indirect benefits from operator (carrier) number portability. For example the Office of Telecommunications (OFTEL) observed:

> As well as the substantial direct benefits (e.g., customers do not have to incur the costs of changing stationery, fewer wrong numbers are dialled), portability provides very significant indirect benefits, assisting greatly in the creation of genuine competition for all categories of customers, driving down prices, encouraging innovation and raising quality.[2]

A more detailed discussion of the benefits of LNP can be found in a Monopolies and Mergers Commission (MMC) Report.[3] Citing a 1992 study by National Economic Research Associates (NERA) for OFTEL, the report identifies three types of benefits.

Type 1 benefits accrue directly to porting customers and consist of:
1. Savings from not having to change stationery and other items (such as advertisements that include a business subscriber's telephone number).
2. The reduced expenses from switching to lower cost operators.
3. The convenience of using the same number and operator for incoming and outgoing calls.

Although the first two benefits are hard to quantify, there is general agreement that the costs are likely to be high enough to have a significant impact on the typical customer's choice of provider. If numbers cannot be ported, new entrants can be expected to focus on selling some outgoing lines and Centrex (or Virtual Private

---

2. OFTEL, "Number Portability: Modifications to Fixed Operators' Licenses," chap. 1, sec. 1.2.

3. Merger and Monopolies Commission, *Telephone Number Portability: A report on a reference under section 13 of the Telecommunications Act of 1984,* Presented to the Director General of Telecommunications, November 1995 (henceforth MMC Report).

Network) services to multiline business customers, who will tend to continue buying their inbound lines from the incumbent local exchange carrier (ILEC). Some residential customers, too, may subscribe to more than one operator's service, using one operator primarily for outbound calls and another for inbound calls.[4] New entrants may also focus on serving the pay-phone market, as most pay-phones are used primarily for outgoing calls. Competition may be vigorous for these market segments, but weak in other segments. With LNP, however, entrants are not constrained to focus on outgoing traffic. They can, if they wish, cater to customers' needs for both incoming and outgoing traffic. Customers will benefit from being able to choose a single operator (from many competitors) who can meet all their needs. The convenience of one-stop shopping is a component of Type 1 benefits. Estimates of Type 1 benefits for the United Kingdom have been vigorously debated, and there is no consensus on their magnitude.[5]

Type 2 benefits accrue to all customers through the enhanced competition facilitated by LNP. The benefits of competition will likely include lower prices for standard "commodity" service. In addition to porting customers who switch operators to obtain a better price, other customers who are attracted by a new entrant's promotional offers may find the incumbent operator willing to meet or beat its rivals' offers. Furthermore, competition is likely to result in a greater variety of products and services as operators seek to differentiate themselves from their rivals in the marketplace. In this process, operators will discover what customers are willing to pay for and will seek to meet their customers' needs. To the extent that they are successful, prices and consumer expenditures on new services may rise, but the value received by consumers will rise even more.[6] On the other hand, as competition lowers prices for subscribers and services that traditionally made disproportionate contributions toward overhead costs, those subscribers who made small or negative contributions to overhead costs may face increased prices or lower quality services. The quantification of Type 2 benefits is difficult, and the

---

4. In the United Kingdom, some residential customers use British Telecommunications plc (BT) for incoming calls and another operator for outgoing calls (MMC Report, para 2.38). Mercury, a new entrant, claims that it could have earned 194 million pounds more over 3 years if it could have carried its large business customers' incoming calls in addition to their outgoing calls (MMC Report, para 2.108).

5. MMC Report, para 2.148.

6. "The ability of end users to retain their telephone numbers when changing service providers gives customers flexibility in the quality, price, and variety of telecommunications services they can choose to purchase. Number portability promotes competition between telecommunications service providers by, among other things, allowing customers to respond to price and service changes without changing their telephone numbers. The resulting competition will benefit all users of telecommunications services." FCC, In the Matter of Telephone Number Portability, Third Report and Order, Released May 12, 1998, at 4 (henceforth FCC Third Report).

MMC views recent estimates for the United Kingdom with some skepticism.[7] However, there appears to be general agreement that the total benefits of LNP are substantially higher than the costs.[8]

Type 3 benefits, according to NERA, accrue to all subscribers in the form of fewer misdialed calls and fewer calls to directory inquiries.[9] Although earlier analyses have attempted to quantify the direct benefits to porting customers and the indirect benefits of the greater competition facilitated by LNP, relatively little is said explicitly about Type 3 benefits in the MMC Report.

For a subscriber who has decided to move or change to another operator, the decision to port his or her number confers benefits on other subscribers. These benefits are apparent if we compare two situations in which a fixed number of subscribers change operators. In the first situation, assume that all customers who change operators port their numbers. In the second situation, assume that the same customers change operators but none port their numbers. For each case, consider the costs incurred by other subscribers. The differences in the costs incurred by the other subscribers in the two situations can be considered an external benefit of porting corresponding to a Type 3 benefit.

These external benefits to nonporting customers of other subscribers' decisions to port include the following:

1. Continued validity of nonporting subscribers' public and private directories and memorized numbers. This benefit will likely be greater for consumers who need to call many other subscribers.
2. Continued validity of programs for speed dialing, selective call forwarding, or other functions implemented in sophisticated customer premises equipment. This includes some international callback services and broadcast fax services, the costs of which would be higher without LNP, as lists of customer numbers and destination fax numbers would need to be updated more often.
3. Reduced telephone operator assistance. When telephone intercepts and announcements fail, as they sometimes do, or lapse, callers turn to operators for help with call completion. This added assistance and the associated expense will be incurred less frequently if LNP is available.
4. Reduction in wrong number and billing disputes. When LNP is not available, the original number is eventually reassigned to a new customer and the customer often receives calls intended for the subscriber previously as-

---

7. MMC Report, para 2.154.

8. For example, the Federal Communications Commission (FCC) concluded that "Although telecommunications carriers, both incumbents and new entrants, must incur costs to implement number portability, the long-term benefits that will follow as number portability gives consumers more competitive options outweighs these costs." FCC Third Report, at 4.

9. MMC Report, para 2.38.

signed to that number. The incremental resource cost of dealing with wrong numbers and related billing disputes is external to the decision to port.

There can be a significant relationship between Type 1 and Type 3 benefits. If a nonporting customer who changes operators acts to ensure that all other subscribers are made aware of his or her new number in a timely fashion, the costs borne by other subscribers will be lower. Thus, customers who switch operators and do not port their numbers can, at some private cost, undertake activities that reduce the transactions costs (such as updating their private directories) to all other subscribers. The distribution of transactions costs across all subscribers will depend on the actions of nonporting customers who switch operators, although one study estimates that the sum of the Type 1 and Type 3 benefits can be quite high.[10]

In addition to the three types of benefits already described, LNP (when implemented on an intelligent network [IN] platform) can also lead to a more efficient use of numbering resources. These *Type 4* benefits are described for the IN approach to LNP adopted in Hong Kong where the Office of Telecommunications Authority (OFTA) has proposed two distinct numbering schemes—directory numbers (DNs) assigned to subscribers and gateway numbers (GNs) assigned to network nodes. In the IN implementation of LNP, a service control point (SCP) maintains a mapping between a DN (associated with a customer) and a GN (associated with an operator). When a customer ports his or her number to a different operator, all operators' SCPs are updated to show the new relationship between his or her DN and a new GN. This arrangement allows for greater efficiency in the use of numbering resources in the following ways:

1. 1. Less churn of DNs. Some subscribers who would change operators even if LNP were not available are likely to retain their DNs with LNP. It follows that the volume of (directory) number changes will be smaller with LNP than without it. As a result the number of DNs removed from circulation before being reassigned to a new subscriber will be smaller, permitting more efficient use of the DN space.

2. Greater scope for GN-to-DN mapping. GNs are assigned to operators and central offices in blocks and are used by switches to route calls efficiently to the proper network nodes. The hierarchical structure of GNs enables switches to route calls using only a few digits in the GN. Different blocks may be used with different densities. Some blocks may be very sparsely used, but it may not be possible to reassign the unused numbers in one block to other blocks, given the hierarchical structure of GNs.[11] If GNs be-

---

10. NERA estimates that these benefits amount to 570 million pounds over 9 years (MMC Report, para 2.39).

11. In the United Kingdom, only about 40% of the numbers assigned to telephone companies in any area are used. See Telecommunications Numbering Policy, OFTEL Statement, 18 October 1995, para 8.

come exhausted, the number of digits in a GN will have to be increased. Because all DNs can be used (they are assigned individually, not in blocks), it may be possible to avoid increasing the number of digits in DNs when GNs are exhausted and must be expanded to include more digits. The benefits to customers and directory providers of retaining existing DNs can be large, particularly if vanity numbers[12] have become popular.

3. The additional costs of "parallel running" during a period of general renumbering, whereby calls are automatically forwarded to new numbers, will be avoided in an IN implementation, as all numbers are "forwarded" from the dialed DN to its assigned GN. Without LNP, DNs and GNs would be the same and exhaustion of GNs would result in changing subscribers' numbers.[13]

Although the broad public benefits of a stable numbering system, the efficient use of numbering resources, and the value to subscribers of a stable directory infrastructure may be hard to quantify, NERA's U.K. study suggests that they may be significant. We stress again that these benefits are largely external to the porting decision, as they accrue to all subscribers, not just those who make the porting decision. The implications of the external benefits for efficient pricing are discussed later.

Costs of Number Portability

There are several technical approaches to implementing LNP. Interim approaches based on remote call forwarding (RCF) and direct inward dialing have been implemented in the United States. In the longer run, most solutions are based on the use of functions provided by the IN platform. Several proposed interim and IN solutions for the United States are described by the FCC.[14]

The United States has settled on a centralized reference database (CRD) architecture where third-party administrators maintain registries with the mapping between GNs and DNs. Competing LECs update their own databases with information obtained from one of several regional CRDs maintained by third-party administrators. The shared costs of these registries can be an important component of LNP costs. Hong Kong has adopted a different approach—a non-CRD architecture—in which each carrier maintains its own database and exchanges updates with each other carrier. In such an architecture, there may be no shared database costs.

---

12. The IN platform can support numerous services in addition to LNP. These include geographic portability "vanity numbers" that spell out the subscriber's name, or have other significance, just as 800 numbers do. Without LNP, subscribers who switch providers can lose a vanity number that they value.

13. This phenomenon is common in the United States when area codes are exhausted.

14. FCC, Notice of Proposed Rulemaking, CC Docket 95-116, July 13, 1995, paras. 55–62 and 36–47.

Incumbent telephone carriers will generally incur costs to condition their networks to support long-term LNP solutions and additional costs that are specifically related to implementing LNP. To the extent that the incumbents continue to be regulated, the recovery of their costs may raise important regulatory issues. Finally, incumbents might be required to pay, on a reciprocal basis, for a portion of the costs of LNP incurred by entrants.

The structure of costs for LNP varies with the technological implementation and specific institutional arrangements selected by the industry. One useful classification by OFTEL has grouped LNP costs into three categories:[15]

- *System setup costs* that result from the decision to implement LNP, whether or not any particular customer ports his or her number. These costs vary with the technology used to implement LNP. RCF technology and IN implementations use different hardware and software, with different implications for costs and cost recovery. RCF is performed by the LEC that lost the porting customer, most often the incumbent. Some IN implementations, such as the location number routing scheme chosen by most U.S. LECs, require the penultimate "N-1" carrier (the originating carrier in a standard local call or the long-distance carrier in a standard toll call) to perform the required translation for call routing. For local calls, the entrant is the N-1 carrier on a large proportion of calls it originates, whereas the ILEC is the N-1 carrier on a very small proportion of the calls it originates.

- *Per line setup costs* are caused directly by a particular customer's request to have his or her number ported. In an RCF implementation, switches must be reconfigured to redirect calls to the new location of the ported customer. In a full IN implementation, each local competitor's SCP database must be updated every time a customer changes operators and ports his or her number.

- *Conveyance cost* is the *additional* cost involved in routing a call to a subscriber with a ported number, as compared to the same subscriber with a nonported number. With RCF, additional switching and signaling resources are used during call setup to a ported number, and additional transmission capacity (voice circuit-miles) are used during the call itself. With a full IN implementation, every call consumes more signaling resources during call setup, but because the call is routed directly to the ported customer, there is no incremental use of voice circuits for the duration of the call.

---

15. MMC Report, p. 4.

These costs can be recovered through a range of end-user charges and interoperator charges. In determining these charges, regulators are often faced with difficult trade-offs, to which we now turn.

REGULATORY PRINCIPLES AND ECONOMIC EFFICIENCY

Cost recovery methods fall in two broad classes: Each operator bears its own costs, or each operator is compensated by other operators for some portion of its costs. When operators are required to bear their own costs, they could pass on the costs to their subscribers in accordance with applicable regulations.

In a network of networks such as the increasingly competitive local telecommunications market, each operator needs to purchase call termination services and possibly other services from every other operator. Every operator is thus a customer of every other operator. Interoperator charges, which are commonly used for call termination, can also be used to recover LNP costs.

The appropriate choice of a cost recovery method will, in general, depend on policy objectives and the specific circumstances of the case. Interoperator charges can potentially be used by one firm to raise its rivals' costs. Charges to subscribers for service termination can potentially be used by one operator to raise customers' costs of switching to a rival operator. In both cases, if these charges are excessive, costs will be (over)recovered and competition will be inefficiently inhibited, but if the charges are too low, costs will not be recovered and inefficient entry will be encouraged. With imperfect information, the regulator faces a difficult trade-off between cost recovery and ensuring that the incentives to entry are right. Several jurisdictions have articulated overarching principles that address these issues. This section provides a brief summary of important cost recovery principles used in the United States, the United Kingdom, and Hong Kong.

OFTEL's Approach

The MMC Report contains a concise description of the six principles for cost recovery suggested by OFTEL.[16] These are:
1. *Cost causation*, which requires a party whose decision causes costs to be incurred to pay for those costs.
2. *Cost minimization*, which requires that all operators and subscribers face incentives to minimize costs.
3. *Distribution of benefits*, which requires that cost recovery schemes should not neglect underlying external benefits.
4. *Effective competition*, which requires that the costs of switching operators not be kept unduly high, and that one operator not have the ability to raise its rivals' costs.

16. MMC Report, pp. 31–38.

5. *Reciprocity and symmetry,* which recognizes that if new entrants also provide LNP to their subscribers, they should be afforded an opportunity to recover their costs in a symmetric and reciprocal manner from their customers and from other operators.
6. *Practicability,* which recognizes that cost recovery for LNP may be simpler if the cost basis and accounting systems for LNP are the same as those used for interconnection charges.

The FCC's Approach

In its First Report on number portability,[17] the FCC enumerated nine performance criteria for long-term number portability measures. Most of these criteria are technical or organizational in nature (at para. 48). The FCC organized its cost recovery principles around the requirement that any cost recovery mechanism for LNP be competitively neutral. This criterion is interpreted to mean that the cost recovery mechanism should not create an incremental cost advantage for any carrier when competing with any other carrier for a specific subscriber (at para. 132). Similarly, the mechanism should not have a disparate impact on the ability of different carriers to earn a normal rate of return on their investment (at para. 135). The FCC stated that in implementing this criterion it will deliberately depart, as a rare exception, from the principle of cost causation, which it has long recognized (at para. 131). Reciprocal compensation arrangements are viewed as a consequence of the requirement of competitive neutrality (at para. 137).

In its Third Report, the FCC adopted many of the tentative conclusions it had reached in the First Report. The costs of third-party administrators will be paid by telecommunications carriers in proportion to their end-user revenues.[18] Shared costs of the database will be allocated in the same proportions. General setup costs and carrier-specific LNP costs (e.g., charges paid to the third-party administrators for querying the database) will be borne by the carriers themselves. ILECs will be permitted to recover these costs through an end-user fee similar to the subscriber line charge, and competitive local exchange carriers (CLECs) will be allowed to recover them in any manner that is consistent with applicable laws and regulations.

OFTA's Approach

OFTA[19] explicitly identifies two compensation principles that govern cost recovery: the *relevant cost principle* (para. 7) and the *principle of cost causality* (para.

---

17. FCC 96-286, In the Matter of Telephone Number Portability, CC Docket No. 95-116, First Report and Order, Released July 2, 1996.

18. Pure wholesalers pay $100 per year to the administrators.

8). The relevant cost principle means that only costs that are directly incurred as a result of the provision of LNP should be recovered through interoperator or customer charges. The principle of cost causality ensures that each operator will pay for the costs it causes all other operators to incur in providing LNP. This principle appears to envision a reciprocal charging regime. In addition, the discussion paper implicitly appeals to other principles used in other jurisdictions, including distribution of benefits (para. 22) and effective competition (para. 2).

ANTICOMPETITIVE EFFECTS AND ECONOMICALLY EFFICIENT PRICE STRUCTURES

LNP, by reducing the switching costs for consumers considering alternative suppliers of local service, increases the prospect of effective local competition. However, LNP offers the incumbent carrier opportunities to engage in anticompetitive behavior.

An ILEC may have an incentive to raise its rivals' costs by inflating interoperator charges for LNP-related services, by imposing high service termination charges on customers who switch to another provider, or by other means such as degrading the accuracy of common databases on which all LECs rely. Moreover, the external benefits and public good nature of directory services, when viewed in isolation, imply that even cost-based prices may result in an inefficiently low volume of porting.

Standard remedies for market failures due to externalities include pricing below marginal cost and creating ownership rights corresponding to the benefits or costs neglected by the market. The major external benefit of LNP is a more stable and efficient numbering scheme and the more stable directory infrastructure it enables. In the United States, this directory infrastructure includes white and yellow pages, other privately published directories, operator-assisted directory services, and a variety of online and offline electronic directories. An ILEC's ownership of significant directory resources enables it to internalize some of the external benefits described earlier and can have an impact on its incentives to engage in anticompetitive behavior and internalize the external benefits discussed earlier.

ILECs' directories generate revenues from many sources. A standard white page listing for a residential subscriber is provided without charge, but vanity listings, unlisted numbers, and yellow page listings generate substantial revenues for ILECs. The prices charged for these services are not the marginal cost of providing the services. Rather, as with other information goods, the price of a listing is based on the value of the listing to the consumer. Thus, the values consumers place on directories is an important determinant of the directory service revenues. Because

---

19. OFTA: "Operator Portability: Recovery of Costs Under the Intelligent Network Approach," Discussion paper, March 14, 1997.

directory publishing (especially the yellow pages) is a profitable business, and because the ILECs are major publishers of yellow pages, an ILEC has an incentive to preserve the stability of its directory listings. Increased stability of the listings increases the value of the directories and the revenues they generate. Thus, ILEC ownership of directories helps internalize the external benefits discussed earlier.

Actions by ILECs that delay or degrade the deployment of LNP may increase the volume of number changes, and thus may reduce the value of ILEC directories and the revenues they generate. ILEC ownership of directories may therefore reduce their incentives to engage in anticompetitive conduct with respect to LNP. However, this reduction in anticompetitive incentives may be reduced if directory charges are regulated and kept low, if ILEC market share in directory provision falls, or if substitutes for directory listings that can be frequently updated at low cost (such as operator-assisted directory services) are more profitable to the ILEC. The anticompetitive incentives of ILECs with respect to LNP cannot be fully understood without quantifying the effects of exclusionary behavior on the margins generated by the full range of directory services offered by ILECs.

CONCLUSIONS

This chapter has analyzed both the benefits and costs of number portability. The benefits of number portability are broad and have significant public good aspects, including a stable directory infrastructure and the lower prices and greater variety brought about by effective local competition. The costs of implementing number portability solutions are likely to be high. In the emerging competitive market, it is important that these costs should not fall disproportionately on the entrant, retarding competition. Neither should the costs of number portability fall disproportionately on the incumbent, encouraging inefficient entry. Regulators have sought to strike the proper balance by developing broad cost recovery principles for LNP. LNP also affects the value of incumbents' directory services (e.g., yellow pages), and the benefits that ILECs derive from more stable directories may help to offset incentives they have to engage in anticompetitive behavior in setting number portability charges.

There are some similarities between telephone number portability and Internet name and number portability. The Internet domain name system maintains mappings between Internet names (such as bu.edu) and Internet addresses (32-bit strings), and this system resembles the mappings between DNs and GNs in LNP databases. However, there are substantial differences in the ownership of directory infrastructures, governance structures, customer behavior, and industry structures of the Internet and the telephone networks. Exploration of these differences is a ripe topic for new research.

6

Promoting Telephone Competition— A Simpler Way

Timothy J. Brennan
University of Maryland, Baltimore

In 1996, Congress passed the Telecommunications Act of 1996 to promote local telephone entry and eliminate any need to keep the regional Bell operating companies (RBOCs) out of long distance. Through late 1998, no RBOC has satisfied the complex and detailed conditions for such entry set out in the Act's Section 271. We propose instead to permit an RBOC to offer long-distance service out of a state if and only if that state deregulates prices for local telephone service. This proposal conforms to the theory of the case, supplies evidence of competition, allows flexible responses to local economics and geographical factors, and is simple. It also raises substantial problems regarding implementation and state government conformity to constituent interests. We consider three potential refinements—creating a separate "loop" subsidiary within the RBOC, permitting long-distance service on a city or exchange area basis, and reconsidering federal regulation of access charges. A proposal to turn to state local service deregulation as the criterion for RBOC entry into long distance need not be perfect to be better than the regulatory policy currently in place.

INTRODUCTION

Since their origin in the wake of the 1984 breakup of AT&T, the RBOCs—the local telephone companies, for most of us—have not been allowed to offer long-distance service to their local telephone customers. In early 1996, Congress passed the Telecommunications Act of 1996 ("Telecom Act," P.L. 104–104, S. 652, 104th Congress) with a great deal of fanfare, in no small part because it promised to promote local telephone entry and eliminate any further need to keep the RBOCs out of long distance.

Were its ideals to be achieved, consumers would reap the benefits of competition for local service. Businesses and households would have a greater opportunity to benefit from economies of scope primarily on the marketing side, encapsulated in the phrase "one-stop shopping,"[1] along with other savings associated with allowing these new entrants and the RBOCs to offer local and long-distance packages. A more controversial claim is that RBOC entry would enhance competition in a currently oligopolistic long-distance service market.[2]

Under the complex framework set out in the Act's Section 271, an RBOC must show that it is offering "just, reasonable, and nondiscriminatory" access and interconnection to its network, so other firms will be able to compete effectively for local telephone customers. Whether or not such competition actually ensues, the RBOC has to satisfy a 14-point checklist of conditions. These deal with interconnecting with entrants and supplying them with "network elements" (e.g., lines to customers' premises, switching services, interoffice trunk lines) that they might need to compete on the proverbial level playing field. The Federal Communications Commission (FCC) then can approve the RBOC's application after it finds that the checklist is satisfied and that granting permission would be "consistent with the public interest."

In no case so far has the FCC found that an RBOC has met the conditions in the Section 271 checklist. Moreover, courts tied up the process. Some decisions currently keep the FCC from conditioning approval on whether the RBOCs adopt the FCC's methods for setting prices for interconnection and these network elements. Another court, albeit since reversed, had found that Congress unconstitutionally named only the RBOCs (as opposed to "all local exchange companies" or "all companies currently subject to antitrust decrees") as those having to meet the Section 271 requirements.[3] Reflecting frustration with this record, Sen. John McCain (R–AZ) introduced legislation in the 105th Congress, S. 1766, to discard Section

---

1. Brennan (1995, esp. 478–480) discussed why the scope economies are probably more on the marketing side than on the technology side. The Telecom Act's required unbundling of network elements presupposes that technological economies of scope within the local exchange are not very substantial. If so, I would not expect substantial functional economies of scope between local and long-distance service.

2. For a recent critique of the state of competition in long distance, see MacAvoy (1998).

3. *SBC Communications, et al. v. FCC* (1997), finding that the Telecom Act was a "bill of attainder"; the 5th Circuit Court of Appeals reversed this ruling in September 1998. The policy discussion takes place in the context of RBOC entry, as they were the subject of the antitrust decree entered into by AT&T and the Department of Justice to settle their antitrust litigation. Accordingly, the RBOCs are the only local telephone companies to which Section 271 of the Telecom Act applies. However, the points made in this chapter apply in principle to any regulated monopoly local telephone company and not just those that were assembled from former AT&T subsidiaries.

271 and lift the restrictions on RBOC provision of in-region long-distance service within a year after its passage.

Perhaps reconsideration is premature; time may smooth these bumps in the road. But with a call for alternatives in the air, I offer here a modest proposal that may meet the policy objectives behind Section 271, while avoiding some of its practical complications: Allow an RBOC to offer long-distance service out of a state if and only if that state deregulates prices for local telephone service.

This proposal has at least four virtues: It conforms to the theory of the original divestiture case, gives states flexibility to respond to local differences in economics and geography, supplies evidence of competition, and is simple. The proposal, however, raises a substantial number of problems regarding implementation and state government conformity to constituent interests.

I begin with an outline of the theory of the government's antitrust case against AT&T, which led to the 1984 divestiture of the local operating companies and the court's imposition and continuation of restrictions on local provision of long-distance service. I then present and examine the Section 271 regulatory substitute to the antitrust decree, which Congress passed and the President signed a dozen years later. The background portion concludes with an assessment of the implementation of Section 271 since enactment of the Telecommunications Act.

With that experience and the theory rationalizing keeping local carriers out of long-distance markets as background, I then turn to an evaluation of the proposal to permit RBOCs to offer long-distance service within a state in its service territory if and only if that state deregulates local service. Following an examination of the rationales and risks inherent in the proposal to leave RBOC diversification to state decisions regarding rate regulation is a brief discussion of three potential refinements. One is the availability of further structural alternatives, in particular, a "LoopCo" proposal to create a separate "loop" subsidiary within the RBOC itself. A second is permitting entry on something geographically more finely cut than the state-by-state basis mandated by Section 271. A third is how federal regulation of access charges affects a proposal to let state deregulatory decisions guide long-distance entry. I conclude with a reminder that a proposal to use state local service deregulation as the criterion for RBOC entry into long distance, or any other proposal, need not be perfect to be better than the regulatory policy currently in place.

THE STORY SO FAR

KeepingLocalTelephoneCompaniesoutofLong-DistanceMarkets

Under normal circumstances, enhancing long-distance competition and one-stop shopping, let alone the presupposition that entrepreneurial vision better guides resource allocation than public management, would justify letting the RBOCs diversify into the closely related long-distance business. However, matters may not be quite so benign when the entrant is a regulated monopolist and the entry is into a closely related market, as economic theory predicts and the antitrust case against AT&T supported (Brennan, 1987). Because the arguments are by now quite familiar, brief summaries should suffice.

DISCRIMINATION

The monopolist provides its regulated service on a delayed or degraded basis to unaffiliated competitors in vertically related markets, creating a competitive advantage from which its unregulated affiliates can profit. In effect, the firm evades regulation by tying the regulated service to an unregulated product and raising the price of the latter to extract the monopoly profits available (but supposedly off-limits) from sales of the former. The specific concern is that an integrated local telephone monopoly would provide inferior access or delays in supplying access connections to its long-distance competitors. A second possibility is that the local telephone company could provide inferior local service to customers that subscribe to a competing long-distance carrier.

CROSS-SUBSIDIZATION

Under rate-of-return regulation, the monopolist can charge (more or less) the average reported cost for its service. Here, the primary concern would be that a local carrier would take personnel, equipment, material, and capital costs incurred to provide long-distance service and designate them as local service costs.[4] If so, it could increase the rates it charges for the regulated service, capturing the profits through effective subsidies to its other businesses. The potential for such subsidies may make credible predatory threats against potential entrants in those markets.

Without monopoly, the local company has no threat to carry out. Attempts to discriminate against unaffiliated long-distance companies or their customers would be futile if those companies or customers could turn to other local service or access providers. In the absence of market power, a local carrier could not fund cross-subsidies, as consumers would turn to local competitors rather than pay higher rates. But regulation is crucial as well. If the local carrier is free to charge the profit-maximizing price, it generally has no incentive to use discrimination as a means for tying long-distance service, or anything else, to its local service so as to raise the latter's price. Cross-subsidization works only if there is a regulator that restricts prices to reported costs.[5]

---

4. As AT&T, MCI-WorldCom, Sprint, and others have sunk substantial amounts of capital into providing long-distance service, they would not appear to be likely victims of predatory actions to drive them out of that market.

Accordingly, the antitrust remedy, exemplified by the AT&T divestiture, was to erect a wall between regulated monopoly local telephony and the provision of competitive services, such as long distance. To keep the problem from recurring, the divested RBOCs were prevented from climbing back over that wall, by excluding them from related lines of business. In 1991, the *U.S. v. AT&T* trial court lifted restrictions on RBOC information service provision. But because AT&T continued to oppose RBOC entry into equipment manufacturing and long-distance service, those restrictions were maintained. The RBOCs had initiated proceedings to vacate these restrictions, but that issue was ultimately resolved not by the courts but by Congress.

Section 271: Tracks, Checklist, Public Interest, and the Booby Trap

Congress did not just vacate the *U.S. v. AT&T* antitrust decree. Rather, in the Telecom Act, it replaced the litigation process with a regulatory procedure relying on a combination of actions by state utility commissions and the FCC, with advice from the Department of Justice. The framework in Section 271 has three components:

TRACKS A AND B

Section 271(c)(1) sets out two broad contexts under which an RBOC may be permitted to offer long-distance service to its in-region customers.[6] The first, so-called Track A route, essentially says that the RBOC must have a state-approved agreement to provide nondiscriminatory access to and interconnection with a competing provider of local telephone service, to businesses and residential customers, that uses its own facilities "exclusively" or "predominantly." Track B offers an alternative for an RBOC in case no competitor comes to the party. The RBOC can still get permission to enter the long-distance business if it has a state-approved access and interconnection offer of terms and conditions that no competitor has either taken or negotiated in good faith to amend. The absence of local competition would not bar RBOC expansion into long-distance markets.

THE "COMPETITIVE CHECKLIST"

Section 271(c)(2)(B) sets out a list of 14 conditions that an RBOC's access and interconnection agreement must meet, whether accepted by a facilities-based

---

5. If regulators can adopt incentive regulation mechanisms such as price caps that divorce rates from costs to give regulated firms an incentive to cut costs, cross-subsidization is no longer a threat (Brennan, 1989).

6. The Act gives an RBOC permission to offer long-distance service out of its region; that is, to customers that are not its local service customers. That they have not done so may speak to the extent to which long-distance prices are substantially above competitive rates.

competitor (Track A) or not (Track B). The list, translated a bit from the Telecom Act's phrasing, is:

1. Equal quality interconnection on nondiscriminatory, cost-based[7] terms.
2. Rights to purchase access to network elements (i.e., physical components of local exchange service) on similarly nondiscriminatory, cost-based terms.
3. Access to poles, ducts, conduits, and rights of way at nondiscriminatory, just, and reasonable rates.
4. Specific availability of local loops running from central offices to customer premises, unbundled from the rest of the network.
5. Specific availability of trunk transmission, unbundled from the rest of the local network.
6. Specific availability of switching, unbundled from the rest of the local network.
7. Nondiscriminatory access to 911, directory assistance, and operator services.
8. Inclusion of customers in "white pages" telephone directories.
9. Nondiscriminatory rights for the competitor to assign telephone numbers to its subscribers.
10. Nondiscriminatory access to signaling systems and databases used for call routing and completion.
11. Number "portability," allowing consumers to keep their phone numbers when they switch local telephone companies, with as little "impairment" of service "as possible."
12. Nondiscriminatory access to whatever a local carrier needs to achieve "local dialing parity," or the ability to connect to long-distance carriers without having to dial extra digits or access codes.
13. Reciprocal payments schedules for terminating calls from competitors to the incumbent's customers (including, but not necessarily requiring, zero-price "bill and keep" arrangements).
14. "Wholesale" service offered for resale at retail prices less costs attributable to marketing, billing, collection, and other "avoided" costs associated with dealing directly with the consumer.

---

7. The term *cost-based* comes from section 252(d)(1) of the Telecom Act. That section says that rates for interconnection and network element charges should be based on cost, and "may include a reasonable profit" (252(d)(1)(B)), yet must be "determined without reference to a rate-of-return" (252(d)(1)(A)(i)). Because "cost-of-service regulation," "rate-of-return regulation," and "just and reasonable rates" are synonymous in regulatory jargon (but apparently not in the Telecom Act), the content of this phrase is elusive. An FCC staffer told me that these provisions were thought to justify its adoption of "total element long-run incremental cost" (TELRIC) rates for the network elements. Perhaps, but only in the formal sense that the sentence "X and not-X implies Y" logically holds for any X and Y.

THE "PUBLIC INTEREST" AND THE BOOBY TRAP

An RBOC believing that it has satisfied the checklist then petitions the FCC for permission to provide long-distance service to its local customers in a particular state. After consultation with the Attorney General (i.e., the Antitrust Division) and the relevant state regulators, the FCC then approves the petition if it finds that the RBOC has in fact satisfied the checklist under either Track A or B (i.e., there is a facilities-based competitor or not) and that "the requested authorization is consistent with the public interest, convenience, and necessity" (271(d)(3)(C)). Those concerned with discrimination and cross-subsidization are quick to cite the "public interest" test as a further guarantee that the FCC can take steps to ensure that competition is protected. For example, the Department of Justice has espoused an open local market standard requiring the incumbent to ensure that local telephone service is "fully and irreversibly opened to competition through all three entry modes envisioned by the Telecommunications Act—facilities-based, resale, and unbundled network elements."[8]

The potential "booby trap" is the immediately following clause in the Telecom Act (271(d)(4)), which says "[t]he [Federal Communications] Commission may not, by rule or otherwise, limit or extend the terms used in the competitive checklist set forth in subsection (c)(2)(B)." In other words, the "public interest" cannot include anything not already identified in the checklist. The extent to which this constrains the FCC's "public interest" determination will likely have to be settled by the courts down the line.[9]

Implementation at the FCC: What Have Been the Roadblocks?

Following the passage of the Telecom Act, a reasonable expectation would have been that the RBOCs would quickly and mechanically satisfy the checklist, most likely under Track B. The controversy about the merits of diversification would then turn on the "public interest" question. Suppose the FCC rejected the application, citing for example the Justice Department's "fully and irreversibly opened to

---

8. Schwartz (1997, at para. 4.) This test includes not just "meaningful implementation of the competitive checklist" (as opposed to merely satisfying its terms?). It goes on to cite the need for "a demonstration that [wholesale systems] are capable of functioning under real world business conditions and of being scaled up appropriately to meet entrant demand," that network element prices "will remain reasonable and cost-based" after long-distance service is authorized, and "the absence of major state or local regulatory barriers or any other barriers that significantly impede competition" (para. 5).

9. For example, if the FCC finds that an RBOC has satisfied the checklist but that there still exist other "major state or local regulatory barriers," will it be allowed to prevent RBOC entry into long distance citing the "public interest," or will it be forced to ignore that circumstance because to do so would add a 15th item to the checklist?

competition through all three entry modes" or "the absence of major state or local regulatory barriers or any other barriers that significantly impede competition" tests. The RBOCs would probably then take the FCC to court, citing the "booby trap" clause of the Telecom Act restricting the FCC's "public interest" criteria to the checklist. If on the other hand the FCC decided that it had to approve the RBOC application because of the "booby trap" restriction to checklist conditions, RBOC entry opponents, most likely the current long-distance service incumbents, would sue, claiming that the FCC had failed to undertake its obligatory "public interest" review.

It has not worked out that way. As of late 1998, no RBOC application has, at least in the eyes of the FCC, succeeded in meeting the Track A and B or checklist conditions. Section 271's appearance of precision seems to belie a reality of considerable interpretive discretion. Some notable examples include:

- In evaluating the initial application from SBC to serve Oklahoma, the FCC found that interconnection agreements with a small provider (20 subscribers at the time) failed to satisfy either Track A or Track B (FCC, 1997a). Track B did not apply because there was an agreement, yet Track A failed because the interconnecting carrier was too small to be deemed competitive.

- Failure to provide operations support systems (OSS) emerged as an important factor in the FCC's evaluations. OSS is defined as "a variety of systems that enable a local telephone company to provide services to its customers such as installation, repair and maintenance, and billing." This may appear to be an addition to the checklist, in violation of the "no extension" clause in (271(d)(4)).[10] However, the FCC warrants its inclusion as part of what it means to provide nondiscriminatory access to network elements, Item 2 in the checklist (FCC, 1997b, at para. 131). As other items in the checklist specify particular network elements (e.g., loops, trunks, and switching), the FCC found that they too imply that OSS is a necessary condition for permitting an RBOC to provide local service.

- In evaluating Ameritech's application to provide long-distance service to its Michigan subscribers, the FCC explicitly stated that it would use "forward-looking economic costs" as the standard for evaluating the reasonableness of network element rates, and found that an incumbent carrier may not unbundle a set of network elements that it itself uses on a combined basis and a competitor wishes to purchase (FCC, 1997c). This is controversial in part because the 8th Circuit Court of Appeals had determined a month before this decision that the FCC exceeded its jurisdiction in prescribing that these pricing rules be included in state-approved inter-

---

10. Whether OSS is in the checklist is apparently a matter of some controversy (Kennard, 1998).

connection agreements under the Telecom Act (*Iowa Utilities Board, et al. v. FCC,* 1997).

- In evaluating BellSouth's South Carolina application, the FCC (and the Justice Department in its reviews[11]) expressed its concern with Bell-South's failure to give its competitors sufficient scope in the terms and conditions by which they can purchase and use network elements on a combined basis (FCC, 1997e).

- The FCC has also raised concerns regarding whether RBOCs offer wholesale customer-specific contract service arrangements (CSAs). The idea here appears to be an RBOC can offer customized services through CSAs under contract rather than tariff to particular subscribers, usually high-volume users with particular needs. According to the FCC, under Item 14 of the checklist, which mandates the wholesale availability of retailed services at avoided rates, the RBOC would have to offer its competitors CSAs at wholesale rates, discounted from retail by avoided costs of retailing. In assessing BellSouth's applications specifically, the FCC claimed that BellSouth failed to do this by offering CSAs to competitors only at original retail rates, and thus failed to meet Item 14 of the checklist (FCC, 1997d, at para. 212, 217).

This record of failure of the tracks and checklists to provide unambiguous guidance for complying with the Telecom Act has precluded an explicit regulatory and legal test between the "public interest" and "nothing beyond the checklist" clauses in the Telecom Act. But the issue has been joined to some degree by explicit pronouncements by the Department of Justice in its advisory capacity and the FCC. The Department of Justice concluded that because SBC had failed to act to make Oklahoma local telephone service "irreversibly open to competition," FCC authorization of SBC's petition would fail the 271(d)(3)(C) "public interest" test (U.S. Department of Justice, 1997a). In the Michigan case, the FCC challenged head-on the "booby trap." It "made clear that the public interest test is an independent requirement that must be met in addition to the other requirements of Section 271, such as compliance with the competitive checklist" (FCC, 1997c). The FCC went on to assert that it has broad discretion to make public interest findings, and that it will consider "a variety of factors" in so doing. The prospect of a court fight over the breadth of the "public interest" standard remains.

STATE DEREGULATION AS THE CRITERION

Reviewing the Context

Developing workable mechanisms to ensure the network externalities and remaining natural monopoly characteristics may inevitably be a slow, complicated pro-

---

11. U.S. Department of Justice (1997b, 1997c, 1998).

cess. However, the paucity of applications and the length and complexity of these evaluations suggests that the process itself may be unwieldy.[12] The emergence of OSSs and CSAs as important FCC desiderata, along with others, suggests that the Section 271 checklist may be capable of indefinite expansion in the depth of its items, if not their number. Simply dropping the ban on RBOC provision of long distance altogether gives no weight to the anticompetitive risks outlined earlier. But, at the same time, treating local telephone service as an effectively permanent natural monopoly, however reasonable when the divestiture was announced in 1982, seems inappropriate today.

FCC Commissioner Michael Powell (1998) proposed retaining but streamlining the Section 271 framework. A first step would be to initiate a collaborative process among the FCC, the Justice Department, state regulators, and carriers (presumably local and long distance) to identify national "best practice" ways RBOCs could satisfy the checklist. These "best practices" would then be tailored to suit specific state contexts, through a similar process undertaken "well in advance" of a specific Section 271 application. With these steps, he suggested, the FCC's review could be undertaken more quickly and with greater expectation of a favorable outcome.

Although this proposal may work, its call for national standards invites the political contentiousness that created an awkward and ambiguous Telecom Act, and the state-specific tailoring could replicate the difficulties associated with the particular reviews as currently undertaken. Accordingly, we should consider a simpler test for allowing local telephone companies to offer their subscribers long-distance service: The FCC should give an RBOC permission to provide long-distance service to its local customers in a state if and only if that state deregulates local service.

Advantages

THE RIGHT FIX

The rationale for keeping the RBOCs out of long distance, following the breakup of AT&T, was to separate competitive businesses, such as long-distance service, from regulated local telephone monopolies. Evading the regulatory profit constraint provides the incentive for local telephone companies to exercise market power indirectly by entering related businesses. As described earlier, they might

---

12. FCC Commissioner Powell (1998) described the process as a "time-restrictive, paper-intensive, adjudicative approach [that] has left our talented and dedicated staff powerless either to give detailed guidance on all aspects of Section 271's competitive checklist or to coordinate as fully with the applicants, the Justice Department and State Commissions as I believe necessary to achieve our collective end: the promotion of competition in local and interexchange markets."

give their affiliated long-distance company a competitive advantage by providing delayed or low-quality connections to their long-distance competitors, or by charging long-distance costs to their local service customers. This theory directly implies that eliminating the regulatory constraint eliminates this anticompetitive incentive for expansion of a local telephone company into long-distance service. Even if state deregulation is premature, a deregulated RBOC could more or less extract monopoly profits from its customers directly without having to diversify. There would be no compelling theoretical reason to presume that diversification by an unregulated RBOC would reduce overall economic welfare.[13]

EVIDENCE OF COMPETITION

Many observers may not share this view that diversification by a monopolist is not automatically a matter for concern. However, the "state deregulation" test should lend them some comfort. A state's decision to deregulate local telephone service should be a good test of whether market forces would adequately promote competitive pricing, further reducing any monopolistic problems with RBOC entry into the long-distance business. RBOC opponents are probably sufficiently organized and able to contest potential deregulatory decisions in states where actual or potential competition may not yet warrant leaving local telephone prices to the market.

REFLECTING DIFFERENT CIRCUMSTANCES

The circumstances under which competition is feasible may differ across states. A state with greater wealth or more local economic dependence on information services may have greater demand for local service, and thus a greater ability to support multiple providers. States also differ in population density, topography, growth, wage rates, education levels, and other factors affecting the ability of new entrants to come into the market at cost sufficient to earn a profit while competing against incumbents. As the authorities closest to the situation and presumably most familiar with the circumstances, state regulators may be better placed to make the decisions as to when local markets are sufficiently competitive to remove regulation.[14] Once regulation is removed, particularly in the face of effective competition, the chief argument for maintaining a ban on local telephone company provision of long-distance service disappears.

---

13. Numerous commentators have suggested that vertical integration by an unregulated local carrier is unambiguously bad if its takeover of a long-distance market would deter future entry into local service. Whatever the merits of such a theory in general, its applicability here is questionable at best. Monopolization of long distance by an RBOC, even to its own customers, seems rather unlikely, particularly if as I propose, RBOC long-distance provision cannot happen unless a state elects to deregulate local telephone service. As discussed next, such deregulation is also a likely indicator of competition in local service.

SIMPLICITY

A 1-point checklist should beat a 14-point one, although simplicity could be illusory, as I discuss next.

Disadvantages

STATES WILL DEREGULATE TOO MUCH

Making long-distance entry conditional on deregulation could lead RBOCs to use their political clout to get deregulation before local markets become reasonably competitive. To some extent, "capture theory" and the economics of public choice suggest that local telephone companies should have disproportionate influence with their regulators. The facts, however, are somewhat ambiguous. Were the regulatory process that profit-driven, RBOCs would have obtained deregulation long ago, back when they really had uncontested monopolies. Persistence of residential subsidies also suggests that the "little guys" possess some electoral clout.

STATES WILL DEREGULATE TOO LITTLE

The opposite contention is that states will never deregulate even if local service becomes competitive. Perhaps state utility commissions will not deregulate themselves into the unemployment line. But there is also ample evidence, most notably from electricity markets, that states will deregulate to reap the benefits of competition, as an economic development measure or a response to the political clout of large buyers (Ando & Palmer, 1998).

DEREGULATION IS TOO DEMANDING

As a rationale, perhaps *ex post*, for the Section 271 approach, sufficient competition to warrant deregulation is too strict a standard; indicia will suffice. Politically that may well be, but the economic rationales are less powerful. All else being equal, it presumes that buyers are more willing to seek out competitors in response to inferior quality than to high prices. The Act may express confidence in regulators to police discrimination, but it is not clear how the criteria in the Act, particularly the checklist, speak to that confidence apart from a generic expectation of

---

14. Without endorsing this argument, Krattenmaker pointed out that this reasoning also implies the Telecom Act's provisions on interconnection agreements and duties in Sections 251 and 252 should also be left to the states. Space limits a full discussion of that proposal, but I tend to agree with it, with the qualifications regarding interstate implications discussed in Section 'Follow-on Deregulation? The Federal Role in Access Charges.'

local competition.[15] A more pertinent justification is that access services long-distance companies need to reach subscribers and callers are more competitive than loop services provided directly to local calling customers. But as discussed later, deregulation of the access services and divestiture of the loops would be a preferable remedy.[16]

BELOW-COST RESIDENTIAL RATES ·

To preserve below-cost residential telephone service, states may be reluctant to deregulate rates. Imposing such requirements on a single competitor, however, is economically and administratively unsustainable under Section 271. A state that wants to promote universal service could do so through an explicit tax and subsidy program, consistent with maintaining efficient and effective local competition in a deregulated environment. Section 102(a) of the Telecom Act envisions a process in which multiple carriers would be eligible to meet universal service goals.

DELAY IN ONE-STOP SHOPPING

Waiting for the states to deregulate could delay restoring the convenience many consumers felt in the days before divestiture when they could meet all of their telephone service needs by dealing with one company. However, one could achieve the benefits of one-stop shopping by allowing resale of local service, prior to allowing the RBOCs into the market directly. The reported difficulties in providing adequate OSS and CSA arrangements to competitors are not encouraging, but these factors are delaying RBOC entry under the current regime as well. Those delays may be a reasonable fact of life; wishing markets were competitive does not make them so.

DE JURE VERSUS DE FACTO DEREGULATION

The most compelling objection may be that identifying deregulation may not be simple. The threat of reregulation of prices may make deregulation illusory, creating incentives to evade any implicit price controls waiting in the political wings. FCC Chairman William Kennard's criticism of long-distance carriers for allegedly not cutting prices in response to reductions in local access charges exemplifies this

---

15. Supporters of the Act could concede that the checklist is irrelevant for preventing discrimination, but should be regarded as the price an RBOC needs to pay to get the "carrot" of long-distance entry. But that carrot seems rather stale if RBOC entry will make the long-distance market competitive, and the RBOCs will no longer be able to profit through discrimination or cross-subsidization.

16. See Section 'The "LoopCo": Relevant Product Markets.'

problem, as has our vacillating "deregulate today, reregulate tomorrow" cable television policy. Another lesson from cable's history could be that shadow regulation is not enough; actual reregulation would be necessary. If so, we may be able to infer de facto deregulation from de jure decisions and policies.

REVERSALS OF FORTUNE

Should reregulation of local service rates force an RBOC out of the long-distance business with all of the upheaval that may entail? Section 271(d)(6)(A) allows the FCC to order compliance, impose penalties, or revoke permissions to provide long-distance service if an RBOC fails to continue to meet conditions for approval. Forcing customers to switch from an RBOC to a competitor, however, is likely to be politically and legally controversial. Antitrust enforcement is an option, albeit cumbersome and costly, if RBOC entry into long distance leads to anticompetitive outcomes. We can only hope that the trend toward increasing local competition is sufficiently inexorable that we will not have to consider such reversals.

POTENTIAL REFINEMENTS

So far, we have been primarily considering whether to let RBOCs as currently configured into the long-distance business on a state-by-state basis, on the basis of state deregulation of local service. The range of issues in the debate about allowing local telephone companies to provide long-distance service suggests some refinements to this perspective.

The "LoopCo": Relevant Product Markets

A different proposal to speed up the provision of some form of integrated local and long-distance service is the idea of a "LoopCo" (Duvall, 1998). Applying the divestiture's rationale of separating competitive from monopoly businesses, the LoopCo idea would redraw the line from its current local–long-distance boundary to one between local loops or other local facilities and retail marketing. The incumbent local company would divest the loops and a few other services, such as 911 and white pages, to create the LoopCo. Rates for these services would continue to be regulated. Undivested switching, bulk transmission, and retail operations could then be deregulated. The incumbent could then offer retail local service and long distance on the same basis as other telephone entrants, all of whom would procure loops at wholesale from the independent LoopCo.

This could be an acceptable alternative to Section 271 or state deregulation as entry criterion. To the extent that a state deregulates a separate corporation that was once part of a local telephone company, that separate corporation should be allowed to provide long-distance service under the standard proposed here. But the specific LoopCo proposal is predicated on the ubiquitous validity of three assertions:

- Efficient integration does not require self-provision of loops.
- Local loops are the remaining natural monopoly in local telephone service.
- Local loops will remain such a monopoly for the reasonably foreseeable future.

None of these assertions seems so compelling to warrant imposing the LoopCo solution as national policy. A virtue of the Telecom Act's flexible approach to network elements is that it leaves open the question of which elements entrants will need to offer efficient local service and integrated telecommunications packages. With cable and electric utility lines already available and with wireless channels increasingly plentiful, the unaided eye might be forgiven for seeing that loops may be the most competitive of the local service markets.[17] In rural areas particularly, scale economies in switching and trunking may well dominate those in loop provision.

Partial State Permissions: Relevant Geographic Markets

The Telecom Act allows entry only on a state-by-state basis, when competitive circumstances can vary considerable within the state. However, urban areas may face competition before rural areas (or vice versa if rural areas are more amenable to entry into loop provision). Central business districts may face competition before it arrives in residential zones. High-income areas may see entry before low-income areas. If an RBOC had a checklist-satisfying agreement with a facilities-based residential and business service provider, but which served only one city in a state, should it provide long-distance service to its customers in the rest of the state? If so, should we be concerned with the risk of anticompetitive discrimination and cross-subsidization affecting customers outside the competitive city? If not, could failure to have a facilities-based entrant in a small rural community suffice to preclude an RBOC's provision of long-distance service throughout the state?

---

17. If policymakers pursue the LoopCo alternative, it becomes more important to consider specifically how the discrimination and cross-subsidization fears play out. Suppose discrimination could be undertaken through schemes involving the provision of bulk access service to long-distance carriers, but not through the provision of loops to end users. Suppose also that there were no inputs into long-distance service that could plausibly be charged to loop cost accounts. Then, there would be no anticompetitive concern with allowing the LoopCo itself to enter the long-distance business. A greater concern would be whether to allow the LoopCo to re-enter other local markets in switching, bulk transmission, and retailing, where the potential for discrimination and cross-subsidization would be greater (Brennan, 1995, pp. 472–474).

The LoopCo speaks to different competitive circumstances among services; these questions follow from different competitive circumstances across regions. By analogy, we could adapt the "state deregulation" criterion to allow the incumbent local carrier to provide long-distance service to regions or sets of customers for whom the state has deregulated local service, with the proviso that the state uses incentive-based regulation for the remaining regulated services or the incumbent divests those regulated businesses (as with the LoopCo). Deregulation should take away most of the relevant incentive or ability to discriminate against long-distance carriers serving customers with deregulated local service. However, the incumbent may still have an ability to shift costs associated with long-distance service, or the deregulated intrastate markets, to the cost base of the regulated businesses. Divestiture would eliminate the ability to cross-subsidize in that fashion; incentive-based regulation would eliminate the incentive.

Follow-on Deregulation? The Federal Role in Access Charges

The primary fly in this theoretical ointment is that the state is not the only relevant regulator. The FCC sets the access charges long-distance carriers pay local telephone companies to originate and terminate calls. Even if a state deregulates local service, FCC regulation of access charges would support, to some degree, a concern with integration by incumbent local telephone companies into the long-distance business. Because the FCC sets access charges using price caps, concern regarding cross-subsidization is somewhat mitigated.[18] However, such firms would retain the incentive to discriminate against unaffiliated long-distance providers.[19]

---

18. One should say only "somewhat." Recent "reforms" of access charges may have been based in part on the idea that local companies were reporting significant profits in access and thus could be subject to rate reductions on that basis. Such reductions based on profit run counter to the idea that rates should be divorced from costs. Because lower costs typically increase profits, any sense in which high profits increase the likelihood of a rate cut re-establishes the tie between costs and rates that incentive-based regulation supposedly breaks.

19. Weisman (1995) argued that access charge regulation need not lead to discrimination, claiming that a vertically integrated local telephone company could make more upstream from access profits by encouraging entry into long distance than it would lose downstream from the added competition. His results, however, hinge on the crucial, related, and debatable assumptions that it is more costly to provide low-quality access to competitors than high-quality access and that the local telephone company has a lower bound on how much it might discriminate against (or refuse to cooperate with) competitors. Without those assumptions, his model is consistent with the proposition that vertical integration reduces the incentive of the regulated incumbent carrier to cooperate with unaffiliated long-distance companies to provide service efficiently.

Accordingly, we need to take federal regulation of local companies into account in implementing the state deregulation criterion. An appealing amendment would be to require that the FCC deregulate access for markets in which states have deregulated local service. It seems reasonable to infer competition in access from competition in local service overall, and if so, automatic follow-on access deregulation would be sensible. Even if that inference is not exact, follow-on access deregulation may remain sensible when compared to the alternative of not allowing local entry unless both the state deregulates local service and the FCC independently elects to deregulate access. The FCC could then reimpose a Section 271-like checklist of its own on its decisions to deregulate access, leaving us pretty much where we find ourselves today.

One might also remove FCC regulation of access altogether, leaving it to the states. Legal impediments associated with the definition of interstate and intrastate jurisdictions would stand in the way, although in an idealistic, theoretical discussion we could dismiss those as well. A more substantial problem is that the effects of state decisions regarding access charges are borne substantially by out-of-state callers who would pay them (Brennan, 1984; Easterbrook, 1983). A state may have an incentive to allow high access charges to extract profits from nonresident callers, perhaps capturing them through taxes imposed on overall local telephone company revenues or profits.

However, that concern applies even if the local telephone market is perfectly competitive, and a state imposed a tax on long-distance access. If targeted at out-of-state callers, such a tax might run afoul of the Constitution.[20] State authority to deregulate access or discriminate against unaffiliated out-of-state long-distance companies could be a tactic by which a state could evade the implicit price constraint imposed by the Constitution. One might hope that the threat of unconstitutional exploitation of nonresident long-distance callers would not in and of itself require perpetual federal regulation of access charges. However, this possibility may justify not letting the decision regarding RBOC entry into long distance be purely a matter for the states without federal authorization based on either state deregulation of local service (as proposed here) or the current Section 271 process.[21]

CONCLUSION

So far, the Telecommunication Act's Section 271 has fostered no successful attempts by RBOCs to provide local and long-distance service. Although this may

---

20. U.S. Constitution, Article I, Section 8 ("... all Duties, Imposts, and Excises shall be uniform throughout the United States").

21. The federal role has not been inconsequential, as state regulators have found their home RBOCs eligible to provide long-distance service. See, for example, Oklahoma Corporation Commission (1997).

well be the appropriate outcome, the complex and apparently open-ended 271 framework of tracks, checklists, and reviews may be part of the problem. Using local service deregulation as the sole criterion for permitting RBOCs to provide in-region long-distance service is relatively simpler as well as theoretically sound, consistent with competitive objectives, and cognizant of variations in the circumstances across states. Moreover, it can be adapted to reflect partial deregulation within states on the basis of services (e.g., following a LoopCo divestiture) or areas (deregulation of local service to urban, business, or high-income regions). It will be more productive if the FCC can and should adopt a policy of deregulating access where states have deregulated local service at the customer end. Measured against the ideal, it has foreseeable problems; but measured against current reality, it may be a reasonably attractive option, meriting further discussion.

ACKNOWLEDGMENTS

Thanks go to John Berresford, Jerry Duvall, Joel Fishkin, George Ford, David Gabel, Evan Kwerel, Tom Krattenmaker, Molly Macauley, David Moore, Robert Pepper, Daniel Shiman, Thomas Spavins, Ingo Vogelsang, and participants in seminars at the FCC's Office of Plans and Policy and the Center for Public Choice at George Mason University for comments. They bear no responsibility for errors, and no agreement with any position taken in this chapter should be inferred.

REFERENCES

Ando, A., & Palmer, K. (1998). *Getting on the map: The political economy of state-level electricity restructuring.* Washington, DC: Resources for the Future.

Brennan, T. (1984). Local government action and antitrust policy: An economic analysis. *Fordham Urban Law Journal, 12,* 405–436.

Brennan, T. (1987). Why regulated firms should be kept out of unregulated markets: Understanding the divestiture in U.S. v. AT&T. *Antitrust Bulletin, 32,* 741–793.

Brennan, T. (1989). Regulating by capping prices. *Journal of Regulatory Economics, 1,* 133–147.

Brennan, T. (1995). Does the theory behind the divestiture in U.S. v. AT&T still apply today? *Antitrust Bulletin, 40,* 455–482.

Duvall, J. (1998). *Entry by electric utilities into regulated telecommunications markets: Implications for public policy.* Washington, DC: Communications Industry Committee, ABA Section of Antitrust Law.

Easterbrook, F. (1983). Antitrust and the economics of federalism. *Journal of Law and Economics, 26,* 23–50.

Federal Communications Commission. (1997a). Memorandum opinion and order, CC Docket No. 97–121. Washington, DC: Author.

Federal Communications Commission. (1997b). Memorandum opinion and order, CC Docket No. 97–137. Washington, DC: Author.

Federal Communications Commission. (1997c). Commission denies Ameritech's application to provide long distance services in Michigan [Press release]. Washington, DC: Author.

Federal Communications Commission. (1997d). Memorandum opinion and order, CC Docket No. 97–208. Washington, DC: Author.

Federal Communications Commission. (1997e). Commission denies BellSouth's application to provide long distance services in South Carolina [Press release]. Washington, DC: Author.

Iowa Utilities Board, et al. v. FCC. (1997). (8th Cir.), No. 96–3321, July 17, 1997.

Kennard, W. (1998). *Statement on Section 271 of the Telecommunications Act of 1996 before the Subcommittee on Communications, Committee on Commerce, Science, and Transportation, United States Senate.* Washington, DC: Federal Communications Commission.

MacAvoy, P. (1998). Testing for competitiveness of markets for long distance telephone services: Competition finally? *Review of Industrial Organization, 13,* 295–319.

Oklahoma Corporation Commission. (1997). Corporation Commission finds Southwestern Bell meets requirements to provide long-distance service in Oklahoma [Press release]. Oklahoma City: Oklahoma Corporation Commission.

Powell, M. (1998). *Wake up call: FCC Commissioner Michael Powell calls for new collaborative approach to Section 271 applications.* Washington, DC: Federal Communications Commission.

SBC Communications, et al. v. FCC and U.S. (1997). Civil Action No. 7:97–CV–163–X.

Schwartz, M. (1997). *Supplemental affidavit on behalf of the U.S. Department of Justice.* Washington, DC: U.S. Department of Justice.

U.S. Department of Justice. (1997a). Justice Department recommends to FCC that it reject SBC Communications' application to provide long distance in Oklahoma [Press release]. Washington, DC: Author.

U.S. Department of Justice. (1997b). Justice Department recommends that FCC deny BellSouth long distance application in South Carolina [Press release]. Washington, DC: Author.

U.S. Department of Justice. (1997c). Justice Department recommends that FCC deny BellSouth long distance application in Louisiana [Press release]. Washington, DC: Author.

U.S. Department of Justice. (1998). Justice Department recommends FCC deny BellSouth's second long distance application in Louisiana [Press release]. Washington, DC: Author.

Weisman, D. (1995). Regulation and the vertically-integrated firm: The case of RBOC entry into long-distance markets. *Journal of Regulatory Economics, 8,* 249–266.

Telecommunication Regulation in the United States and Europe: The Case for Centralized Authority

William Lehr
*Columbia University
and Massachusetts Institute of Technology*

Thomas Kiessling
Harvard University

This chapter makes the case for a centralized regulatory authority to realize the twin goals of telecommunications liberalization and promotion of integrated infrastructure. In Europe, the debate focuses on the interpretation of the subsidiarity principle and the appropriate regulatory roles for member states and the European Commission (EC); in the United States, the conflict is between state regulatory commissions and the Federal Communications Commission (FCC). We explain how the weakening of centralized authority in favor of state-level regulators (public utility commissions [PUCs] in the United States and national regulatory authorities [NRAs] in Europe) favors incumbent carriers (i.e., Bell operating companies [BOCs] in the United States and telecommunication organizations [TOs] in Europe) by creating multiple veto points and raising the costs of implementing procompetitive supply provisions. A case in point is the recent legislative setback for the FCC (i.e., 8th Circuit Decision, currently under appeal). We demonstrate the need for centralized regulation in the areas of interconnection pricing and licensing policies. Moreover, we explain how the growth of the Internet accentuates the arbitrariness of the state–interstate (national–international) distinction.

INTRODUCTION

The twin goals of telecommunications liberalization and promotion of integrated infrastructure require a centralized regulatory authority; however, concerns over local autonomy conflict with this need. In Europe, the debate focuses on the allocation of jurisdiction between NRAs in the member states and the EC; in the United States, the conflict is between state PUCs and the FCC. Although the tension between local and national regulatory institutions is not new, the issue is both more important and more difficult to resolve today.

First, a centralized regulatory authority is needed today if efforts to promote increased local competition and deregulation (in the United States) or liberalization (in the European Union [EU]) are to be successful. The policy challenge is to manage the transition from monopoly regulation of a dominant incumbent carrier to a competitive market with a level playing field for both the incumbent and new entrants. Creating this level playing field means eliminating both regulatory and economic barriers to entry. When most of the strongest potential competitors to the incumbent operate in multiple local jurisdictions, heterogeneous local rules tilt the field in favor of the status quo and the dominant incumbent local carrier. In the United States, this favors the incumbent local exchange carriers (ILECs) such as Bell Atlantic, SBC, or US West. In Europe, it favors the national incumbent operators (TOs) in the EU such as France Telecom or Deutsche Telekom. An ILEC or TO can take advantage of heterogeneous rules and multiple regulatory fora to deter or delay increased competition. A centralized regulatory authority can help minimize opportunities for such behavior.

Second, a strong centralized authority is needed to facilitate deregulation. It is preferable to roll up the regulatory carpet from the edges. The process of liberalization is likely to proceed more rapidly and will be easier to manage and coordinate if authority is centralized first. On the other hand, if the centralized authority is eliminated first, there is a significant risk that local deregulation will proceed asymmetrically, if at all.

Third, the emergence of the Internet and the goal of promoting an integrated global information infrastructure reduce the validity of assigning regulatory jurisdiction based on geographic boundaries. The Internet is inherently footloose, increasing the difficulty of asserting local control. Allocations of jurisdiction on the basis of intrastate–interstate (in the United States) or national (in Europe) boundaries made more sense in a telephone-only world, but are no longer sensible in the Internet age. Attempts to apply asymmetric local regulations may prove futile, but they may also distort or deter investment that is needed if the Internet is to continue to grow and evolve.

The need for a strong centralized authority may be greater and prospects for satisfying this need are dimmer, largely for political rather than economic reasons. In the United States, the FCC's ability to serve effectively as the centralized authority has been called into question by a series of decisions by the 8th U.S. Circuit Court of Appeals (8th Circuit). In Europe, there is no such thing as a Euro-FCC and creating one in the present political environment is likely to be extremely difficult. In both the United States and Europe, strengthening or creating an effective centralized regulatory authority will require overcoming significant legal and institutional challenges. In this chapter, we do not address these issues, focusing instead on presenting the economic arguments for why a weak or nonexistent central regulatory authority is detrimental to promoting competition and liberalization and is more harmful today than in the past.

ECONOMICS OF DUAL REGULATION

Both the United States and EU have dual regulatory systems consisting of local regulatory authorities and a centralized authority. In this section we examine the economic basis for allocating jurisdictional authority, offering two arguments in favor of (and one against) centralizing authority, as follows:

- Coordination and spillover externalities: *yes,* especially now with Internet.
- Local information and participation: *no,* more important in the EU than in the United States.
- Regulatory costs: *yes,* even more so with deregulation and in light of rent-seeking costs (regulatory capture), which is especially relevant in the EU.

The following subsections explain these arguments in greater detail.

Coordination and Spillover Externalities

When there are spillover, coordination, or network externalities across multiple local domains, centralizing authority offers an obvious mechanism for assuring that these are appropriately internalized[1] In the case of telecommunications networks there are substantial externalities because the same facilities are used to support both local and interstate or cross-border services.

---

1. Coordination, spillover, and network externalities are common in telecommunications. Coordination externalities arise when activities in one domain need to be coordinated with activities in another domain. For example, a telecommunications service provider that provides service in multiple states would need to coordinate facilities planning for its backbone network that is shared by each of the states. Spillover externalities occur when activities in one domain produce costs or benefits in another. For example, mass media advertising is likely to spill over to adjacent markets. Network externalities arise because the value subscribers place on network access is usually increasing in the size of the total subscribership (i.e., telephone callers value telephone service more when they can call more people). Network externalities make a larger network more valuable than a smaller one.

The value of centralized authority increases with the degree of market fragmentation across geographical submarkets (EU member states, U.S. states). The EU in particular has traditionally been characterized by substantial market fragmentation. Lack of service standardization, widely differing supply conditions, and the unavailability of many cross-border services are leading to large welfare losses.

The externalities and spillovers are more apparent at the wholesale level (between carriers) than at the retail level (services sold to end-users) when competing suppliers are active in multiple local markets (which is particularly relevant in the case of U.S. ILECs). In that case, heterogeneous regulations may distort investment incentives or operating behavior, as carriers are encouraged to venue shop or otherwise arbitrage regulatory distortions.

Local Information and Participation

There are two important reasons for decentralizing authority. First, decentralizing authority may be advisable to take flexible account of differences in local circumstances and to economize on information costs. For example, the costs of building a local telephone network are different in the mountains of Colorado and the plains of Kansas. In the EU, the differences are less a matter of construction costs than of different institutional, cultural, and economic legacies.

Decentralization may also be advisable if information is most efficiently collected and maintained locally. For example, effective regulation of local incumbents requires collecting significant amounts of data. Local authorities may be in a better position to gather and synthesize this information. However, as we explain later, decentralized information management becomes more problematic during liberalization and when the incumbents are active in multiple local markets (i.e., the information is no longer local).

A second related reason for decentralizing authority is facilitation of local participation. For telecommunications, this is most important with respect to issues of particular local concern such as the retail-level pricing of local services and the quality of local customer service. Local oversight of these issues may be justified on these grounds. On the other hand, centralization may lower participation costs for issues that affect multiple domains. For example, issues that concern carrier competition affect multiple local jurisdictions and require an understanding of technical, regulatory, and economic issues that may not be readily available locally.

Regulatory Costs

The costs of regulation affect an assessment of the appropriate level of centralization in three ways. First, to the extent that local authorities confront similar problems that result in similar decisions, centralization may reduce the administrative costs of duplicate regulation. In principle, these benefits could also be realized by allocating responsibilities among specific local authorities, however, this would not reduce the shared and common costs of maintaining multiple local authorities. These costs may be increase as the regulatory challenge becomes more complex and requires more specialized and expensive human capital resources and the funds available to sustain such resources become more scarce. For this reason, liberalization and industry convergence are likely to increase the need to centralize authority.

Second, when regulators confront an environment of great uncertainty, there are advantages to experimentation. Decentralization of authority that allows flexible heterogeneity in approaches may be useful in discovering the best policy approach. This is sometimes referred to as the *laboratory of the states*.[2] Although this may prove very useful, a strong centralized authority is desirable when it comes time to disseminate and implement the optimal solution to overcome resistance from laggard local authorities. In the case of promoting local competition in the United States, the laboratory experiments were run for over a decade; with passage of the Telecommunications Act of 1996 it was time to implement the national solution. In the case of the Internet, we do not yet know how these markets will evolve, so regulation seems premature at both the local and centralized level.

Third, *ceteris paribus*, decentralized regulatory authority is likely to be more cumbersome than centralized authority, making it more difficult to change the status quo. This is desirable when there is a risk of regulatory capture by a narrow interest group. It is not desirable when the goal of policy is to change the status quo. This is the case with respect to promoting liberalization and increased competition. Overall, therefore, the economics of regulation suggest that increased centralization is desirable.

DUAL REGULATION IN THE UNITED STATES AND THE EU

As noted earlier, both the United States and EU have dual regulatory systems. In both cases, there has been a trend toward increasing centralization, although the United States has progressed substantially further. In the United States, there has been a presumption that the central authority has a right to preempt local authority, with the burden of proof being on the local authorities to demonstrate that such preemption is not appropriate. In the EU, the subsidiarity principle embodied in the EC constitution,[3] implies the opposite approach: There is a presumption that

---

2. See Noll and Smart (1989).

3. This principle is embodied in a number of provisions of the EC Treaty, for example in the EU antitrust legislation Art. 85 and 86 EC Treaty. These rules only apply to member states if cross-border trade is impacted to a considerable extent. If this is not the case, member states' antitrust rules apply instead. For further discussion of the regulatory landscape in the EU and the role of the subsidiarity principle, see Kiessling and Blondeel (1998).

authority resides at the local level, with the burden of proof being on centralized authorities to justify their role. As we explain later, we advocate stronger centralized authority in both cases, but these alternative approaches are appropriate to the differing circumstances in the United States and the EU.

Dual Regulation in the United States

In this section we briefly review the roles of the main regulatory actors in the United States: the FCC and the state-level PUCs[4].

Historically, the PUCs have been responsible for regulating intrastate telecommunications services, and the FCC has been responsible for interstate services.[5] This demarcation of responsibilities has always been somewhat arbitrary because the same facilities that support local calling services also provide access to interstate toll services. Over time the FCC extended its authority by asserting its right to preempt local authorities on issues related to interstate services.[6] For example, the FCC forced the opening of the customer premise equipment (CPE) market to competition and deregulated enhanced services in its Computer II decision in 1980, over the opposition of state commissions.

More recently, the FCC's authority has been called into question by the decision of the Eighth U.S. District Court of Appeals to strike down a portion of the FCC's Interconnection Order, which the FCC issued as part of its effort to implement the Telecommunications Act of 1996's procompetitive rules for opening and unbundling the local access networks to competing carriers.

Dual Regulation in the EU

Dual regulation emerged somewhat later in the EU than in the United States. In Europe, telecommunications were regulated exclusively at the member state level until the early 1980s. The Commission applied its competition policy to telecommunications for the first time in 1985[7] and in 1987 presented a framework for future regulation and liberalization in the telecommunications sector.[8] In contrast to the United States, still today there is no designated, central (EU-level) regulatory body in telecommunications. Regulatory policy is conducted in parallel by several, relatively independent policymaking authorities that often pursue conflicting

---

4. For a more complete discussion, see Vogelsang (1994), Kellogg, Thorne, and Huber (1992), or Noll (1989).

5. Section 2 of the Communications Act of 1934 limits the responsibility of the FCC to interstate and international telecommunications.

6. See Vogelsang (1994).

7. In a landmark decision the Commission found in 1985 that British Telecom had abused its dominant position in the telecommunications market (see Ravaioli, 1991).

8. Commission of the European Communities (1987).

goals. However, the Commission has been attempting to impose itself as the de facto EU-level regulator in telecommunications. In what follows, we summarize the current scope of dual regulation between the member states and NRAs on the one hand and the EU institutions (Commission, Council, and Parliament) on the other hand.

THE EUROPEAN COMMISSION—DIRECTORATE GENERAL IV (DGIV; COMPETITION)

DGIV is responsible for EU competition policy. DGIV is the main architect of the Commission's liberalization policy in telecommunications and its central instrument has been Art. 90 EC Treaty, which has been used to liberalize telecommunications markets (e.g., services other than voice telephony in July 1990, voice telephony and infrastructure provisioning in January 1998, etc.[9]

THE EUROPEAN COMMISSION—DIRECTORATE GENERAL XIII
(TELECOMMUNICATIONS, INFORMATION MARKET, AND EXPLOITATION OF
RESEARCH)

DGXIII is responsible for the execution of the EU research and development programs in telecommunications, the Open Network Provision (ONP) legislation, and various harmonization and standardization measures. DGXIII also played an important role in the definition of the core regulatory competition framework. The draft process of both the 1997 Interconnection Directive[10] and the 1997 Licensing Directive[11] was driven by DGXIII.

COUNCIL OF THE EUROPEAN UNION

The Council of Ministers is comprised of the Ministers of member states that are responsible for telecommunications policy, and therefore represents the member states' interests. Regulatory measures of the Council often express political compromises between the member states. The Commission depends crucially on support of its liberalization measures from the Council. The Council and the EU Parliament have passed the core regulatory framework enabling the transition to

---

9. For a succinct history of telecommunications liberalization in the EU see Kiessling and Blondeel (1998).

10. The European Parliament and Council of the European Union, Directive 97/33/EC of 30 June 1997 on interconnection in telecommunications with regard to ensuring universal service and interoperability through application of the principles of ONP. *OJ L 199/32 (97/33/EC, 26.7.97)*, 1997.

11. The European Parliament and Council of the European Union, Directive 97/13/EC of 10 April 1997 on a common framework for general authorizations and individual licenses in the field of telecommunications services. *OJ L 117/15 (97/13/EC, 7.5.97)*, 1997.

competitive markets in telecommunications (i.e., the Licensing Directive[12] and the ONP Interconnection Directive.[13] However, the Council has also blocked many measures proposed by the Commission in, for example, the areas of market entry liberalization, licensing, and so on, thereby expressing the opinion of conservative member states.

MEMBER STATES AND NRAS

The central objective of member states is to control the evolving national regulatory and market environment. It is therefore in the interest of member states to keep the Commission from extending its regulatory powers into areas the member states consider to be under national regulatory responsibility[14] As a result, the NRAs are currently working to impose themselves as the prime regulatory authorities for the transition toward competitive markets.

THE NEED FOR A CENTRALIZED AUTHORITY

In the introduction to this chapter, we offered three reasons a centralized regulatory authority is more important today. These included the promotion of local competition in the face of resistance from an entrenched incumbent, more efficient management of overall deregulation, and the changes in networks implied by the emergence of the Internet. In the following three subsections, we explore each of these arguments in greater length.

Promoting Competition. A strong centralized authority is needed to promote telecommunications competition. The biggest challenge facing policymakers in the United States and in the EU is how to promote efficient competition for local services, which remain a de facto monopoly virtually everywhere. Heretofore, the economics and the regulatory legacy have protected the dominant position of the incumbent carrier. In the past, most analysts believed that provisioning telecommunications networks was a natural monopoly (either because of network interconnection externalities or scale and scope economies). This helped justify regulating telecommunications as a protected monopoly. In most of Europe, the telecommunications provider was publicly owned; in the United States, the Bell System was private, but was subject to comprehensive regulatory oversight. With changes in the market and technology, it became feasible to introduce increased amounts of competition along the telecommunications value chain. Thus, recent regulatory efforts have rightly concentrated on introducing competition in the re-

---

12. The European Parliament and Council of the European Union op cit., Ref 111.
13. The European Parliament and Council of the European Union op cit., Ref 100.
14. Analysys, *Network Europe: Telecoms Policy to 2000.* Analysys Publications, Cambridge, England, 1994, pp. 8ff.

maining monopoly areas (i.e., local services in the United States and local as well as long-distance services in the EU).

Introducing local competition requires a change in the regulatory paradigm. Regulators need to remove regulatory and economic barriers that deter competition from other carriers. Instead of protecting the regulated incumbent's market from cream-skimming entry, the regulator must develop policies to promote the emergence of competition. The dominant incumbent carrier has little incentive to cede market share to entrants willingly. By defending the status quo and resisting the implementation of new policies, the incumbent can forestall the implementation of market-opening, procompetitive regulatory reform.

Examples of how incumbents may exploit dual regulatory regimes to slow the progress of competition abound. In the United States, following more than a decade of state-level experiments in alternative regulatory regimes (i.e., the laboratory of the states), Congress passed the Telecommunications Act of 1996. Passage of this act signaled general recognition that a national policy was needed for local competition to succeed, yet almost 3 years later, the Act has still not been successfully implemented anywhere. Similar issues are debated state by state, and it does not even matter if the states all decide identically on the same issues, as is often the case[15] Arguing the same contract provisions between the same parties with often the same expert witnesses in state after state serves only to slow the process of implementing the Act.

In addition to delay, heterogeneous entry rules create entry barriers for competitors who compete in multiple local areas. In the United States, the ILECs operate in multiple states, as do most of their competitors. Requiring these competitors to develop state-specific infrastructure provisioning and marketing plans increases entry costs. The regulatory uncertainty and the staggered sequence of procedural decisions also contribute to higher entry costs.

In the EU as well, without the EC as a central regulatory driving force, market competition would have been further delayed and more fragmented due to resistance from conservative member states and national dominant network operators. The following summary of major events on the road to liberalization illustrates this point:

- May 1992: The Council refuses the Commission's proposal to rapidly eliminate the remaining monopolies. In its decision the Council expressed the will of the majority of member states.[16]

---

15. For example, in each of the 14 states in which US West is the ILEC, US West has argued that it should not be required to comply with the FCC's interconnection order (see *First Report and Order, In the Matter of Implementation of Local Competition Provisions in the Telecommunications Act of 1996*). In each state, the PUCs have eventually upheld substantial portions of the Order. These include such things as requiring US West to permit resale of all services, unbundling at least the set of elements identified in the FCC's order, and implementing electronic interfaces at parity.

- April 1993: The Commission's proposal to liberalize cross-border telephony services in the EU on January 1, 1996 fails to gain support from member states.[17]
- July 1993: The Council confirms January 1, 1998 as the date for the full liberalization of all remaining monopolies. This date had been proposed by member states.[18]

Efficient Liberalization and Deregulation

A centralized authority is needed to coordinate and manage telecommunications deregulation. Lack of coordination among local authorities in the pace and way in which deregulation proceeds may result in heterogeneous rules that will distort competition and incentives to invest or comply with regulations. Disparate regulatory regimes create opportunities for venue shopping whereby firms whose activities are regulated in one market may seek to move those activities to another, less regulated market. This makes it more difficult to enforce remaining regulations and raises the costs to competitors active in multiple markets.

In addition, as liberalization proceeds, regulators will relinquish resources and relax requirements for information sharing. This will reduce the regulators' capability to regulate at the same time that competition and convergence will be fueling the rise of increasingly complex supplier relationships and organizational forms. In this environment, scale and scope economies are likely to make it more efficient to concentrate regulatory expertise in the central authority.

The need for a centralized authority is perhaps best understood if one considers the alternative: deregulating from the center outward. If followed to its conclusion, we may end up with local authorities intact, but no centralized agency capable of coordinating decisions, sharing information, and economizing on duplicative efforts. In this case, it will be even more difficult to effect policy reforms to the status quo.

Maintaining or increasing the power of a centralized authority is not inconsistent with rapid deregulation. Once local regulations have been relaxed and competition is firmly established, it will be possible to deregulate at the center as well.

The Internet and Geographic Boundaries

The emergence of a global communications infrastructure, as exemplified by the Internet, increases the benefits of centralized versus local regulation. This is due

---

16. Telecom Markets, 1992, 25 June 1992, 1–2.

17. See Schenker (1993).

18. Council of the European Union, Resolution of 22 July 1993 on the review of the situation in the telecommunications sector and the need for further development in that market. *OJ C 213/1 (93/C 213/01)*, 1993.

to a number of factors, including changes in market structure, regulatory approaches, and the technology of the Internet.

With globalization and industry convergence, the potential spillover effects or externalities associated with the telecommunications sector have increased substantially.[19] A global communication infrastructure reduces transportation costs, breaking down geographic boundaries between markets. Consumers and potential suppliers may more easily collect and share information about product offerings and prices. The Internet reduces the entry costs for local retailers interested in participating in wider markets, or of national and global retailers participating in local markets. This is true of the communication services themselves, as well as the trade that they support.

Industry convergence also poses important challenges for regulatory policies in other domains such as content, privacy, intellectual property, tax policy, and security—all issues that require national (in the United States) or EU-wide oversight. More traditional aspects of regulatory policy such as cost separations by markets or services are much more difficult in a world of converging infrastructure. For example, in the United States, the allocation of costs to interstate and local markets or between regulated and enhanced services becomes increasingly arbitrary because firms are using common or shared facilities to compete in multiple services (e.g., service bundling to offer one-stop shopping or integration of local, long-distance, and international services) and because of changes in the technology (e.g., packet switching). The increased complexity and arbitrariness of cost allocation procedures makes it more difficult and error-prone to sustain demarcations of regulatory authority based on geographic boundaries.

It is also important to understand how the emergence of the Internet as a new networking paradigm reduces the relevance of geographic boundaries, thereby enhancing the need for centralized authority. First, the basic features of the Internet make it less amenable to local regulation:

- *Packet switched, not circuit switched*: Increased routing options and less hierarchical switching increase the extent to which local and interstate or EU-wide facilities are shared or common.
- *End-user control*: With network intelligence shifted to the periphery, it is less feasible to sustain arbitrary regulatory-mandated heterogeneity at interconnection points in the backbone (i.e., across state or national borders); the boundary between CPE and the network is blurred.

---

19. Convergence of the computer, data communications, and telecommunications industries on the network side; convergence of entertainment media, publishing, and interactive multimedia services on the content side; and integration of local, national, and global markets increase the potential for spillovers across industry, technology, and market boundaries relative to the earlier world of plain old telephone service and separate networks for television distribution, data communications, and telephony.

- *Multimedia*: On the Internet, traffic is multimedia (voice, video, data) and hence much more heterogeneous (with respect to value, source of origin—receiver or sender). This makes it more difficult to develop an appropriate basis for metering traffic to establish prices or allocate costs.
- *Open, interoperable standards*: Encourage interconnection of existing diverse infrastructure—further increasing spillover effects. Heterogeneous local regulation that affects the evolution of Internet technology (e.g., local filtering requirements required to be implemented in router software) poses a significant risk for the continued evolution of the Internet.
- *Internet historically not regulated*: The Internet has been subject to substantially less regulation than the incumbent telephony carriers. If the Internet evolves into the primary platform for our global communications infrastructure—supporting telephony as one application among many—then it will be subject to communications policy. Implementation of a coherent policy will be hindered if there is a legacy of disparate local regulatory policies that must be rationalized and if there is no strong centralized authority.

THE UNITED STATES AND THE EU EXPERIENCES DIFFER

Although similar in many respects, there are important differences between the EU and the United States that make the need for centralized authority less important in the EU. To put it differently, central authority in the EU should fulfill a more circumscribed role.

The United States and the EU are obviously two economic areas with very different economic and political characteristics. The United States shares a common language, culture, and with minor differences, set of political and regulatory institutions. In the EU, national differences are substantially more pronounced, with language being only the most obvious distinction. These differences make the case for centralized authority categorically different than in the United States. Although the extent of cross-border telecommunications demand in the EU is comparable to interstate demand in the United States, the supply side has been historically fragmented into national markets. Although this in itself bolsters the argument for centralized authority, the resulting fragmentation in supply promotes nationally oriented constituencies and thus strong local regulation. Only recently have the dominant national operators in the EU such as British Telecom and France Telecom begun significant efforts to offer services outside of their home countries, either directly or through strategic alliances.

Differences in regulatory market models in the EU provide another reason the need for a centralized regulatory authority in the EU is less strong than in the United States. The United States is by and large characterized by more homogeneity of views as to the basic competitive framework. This has been further enforced by the Telecommunications Act of 1996. In contrast, there is no general agreement in the EU on how best to promote competition. For example some EU countries strongly promote facilities-based infrastructure-based competition, whereas others put the emphasis on service-based competition. Moreover, even the countries that are seeking to promote infrastructure competition differ with respect to the appropriate mechanisms for facilitating new network investment.

An example of how this balance might be achieved is provided by the experience of the EC with respect to the subject of carrier preselection. Since the early 1990s, the U.K. government had encouraged the construction of competitive local access infrastructure by giving local operators certain market advantages. These include allowing new access operators to "own" the customer (i.e., the access operator receives all revenue from the end-to-end call and controls how its subscribers' calls get routed in the long-distance and the termination network). The new access carriers argued that carrier preselection would reduce their profit margins because the customer now controls the choice of the long-distance operator and the latter will bill the customer directly. The new providers therefore argue—supported by Oftel, the U.K. regulator—that carrier preselection would endanger the viability of investment in competitive local infrastructure[20].

The EC's Draft Directive on Operator Number Portability and Carrier Pre-Selection of January 1998 includes the obligation of local access providers that command significant market power to implement carrier preselection. Market experiences in the United States and Australia show that this helps bring down long-distance tariffs and introduce customer choice. However, no obligation was imposed in the Draft Directive on access providers that do not command significant market power to offer carrier selection. This effectively addresses the United Kingdom's objections against carrier preselection, leaving it up to other member states to oblige carrier preselection on all carriers if they wish to do so.

In this case, the central regulatory authority's mandate is limited to a regulatory principle for which consensus can be reached between the member states: the imposition of carrier preselection on local access providers that command significant market power. This provision is compatible with proinfrastructure policies pursued by member states like the United Kingdom.

CONCLUSIONS

On both sides of the Atlantic, communications policymakers are seeking to promote competition and liberalization, while assuring the provision of an integrated, global, communications infrastructure. Realization of these goals requires a strong centralized regulatory authority. Unfortunately, in both the United States and Europe, this authority is inadequate. In the United States, the FCC's authority has

---

20. See Molony (1997).

been challenged by a series of decisions from the 8th Circuit; in Europe, there is no effective EC-level regulator.

This chapter examines the economics of dual regulation and the history of this system in Europe and the United States and seeks to make the case for a strong centralized authority. The need for such authority is especially important in light of industry convergence and the growth of the Internet.

With convergence, communications networks are becoming increasingly integrated with respect to the types of traffic handled, the types of facilities that support that traffic, and the geographic markets in which carriers participate. This increases the potential for spillover and coordination externalities, thereby increasing the risk and costs that heterogeneous local regulations will harm incentives for efficient infrastructure investment and service provisioning. Strong centralized authority is needed to address these risks and help internalize these externalities.

With liberalization, the ruling regulatory paradigm is to promote competition wherever possible. This poses a substantial threat to the dominant position of incumbent carriers and provides them with a vested interest in protecting the status quo regulatory and market environment. Complex and heterogeneous dual regulation creates multiple veto points that are vulnerable to strategic exploitation by an incumbent wishing to forestall regulatory reform or to increase rivals' costs. This provides another important reason for providing strong centralized regulatory oversight over communications policy. If competition is to be successful, the centralized authority should have effective jurisdiction over issues related to the basic structure of competition.

Although these arguments apply on both sides of the Atlantic, it is obvious that the states that comprise the United States are significantly more homogeneous and more integrated than the member states of the EU. These differences imply that the jurisdiction and power of a centralized authority should be much more circumscribed in Europe than the United States. Nevertheless, in both regions, the status quo needs to be revised in favor of stronger centralized authority.

For the EU, we recommend transferring considerable responsibility from the NRAs to an EU-level regulator[21] This regulator could be situated within the EC or established as an independent European Regulatory Authority (ERA) in telecommunications. The Commission will examine the need to set up an ERA as part of the EU Sector review in 1999. To gain support from the member states for an ERA that is vested with the necessary statutory powers, the ERA should be established as a Commission of member state NRA representatives. This would ensure that member states keep sufficient control of the ERA's EU-wide regulatory policies and that the NRAs' hands-on experience in national regulation is duly considered by the EU-level regulator.

---

21. This viewpoint has been expressed by Commission officials and policy observers alike (see Public Network Europe, 1997; Espicom, 1997).

For the United States, we recommend that the FCC's ability to preempt state regulatory authorities with respect to communications policy be reaffirmed and extended, especially regarding issues directly related to the promotion of local competition and the implementation of the procompetitive provisions of the Telecommunications Act of 1996. On economic and policy grounds, we disagree with the position of the 8th Circuit and hope that these decisions will be overturned by the Supreme Court when it considers these issues sometime in 1999. Irrespective of whether one would like to see more or less telecom regulation in the United States, we think it is important that the FCC's authority be maintained until such time as deregulation is more advanced at the state level.

ACKNOWLEDGMENTS

William Lehr would like to acknowledge the support of the MIT Internet & Telecoms Convergence Consortium. In addition, many of the ideas presented here were developed in discussions with my colleague Glenn Hubbard at Columbia University. Dr. Lehr has provided expert testimony on behalf of IXCs in regulatory proceedings in the United States.

REFERENCES

Commission of the European Communities. (1987). *Green Paper on the Development of the Common Market for Telecommunications Services and Equipment* (COM87). Brussels, Belgium: Author.

Espicom. (1997). *European telecoms regulation: Leveling the playing field.* London: Espicom Business Intelligence Publications.

Federal Communications Commission. (1996). *First Report and Order, In the Matter of Implementation of Local Competition Provisions in the Telecommunications Act of 1996* (CC Docket No. 96–98). Washington, DC: Author.

Kellogg, M., Thorne, J., & Huber, P. (1992). *Federal telecommunications law.* Boston: Little, Brown.

Kiessling, T., & Blondeel, Y. (1998). The EU regulatory framework in telecommunications: A critical analysis. *Telecommunications Policy, 22(7)*, 571-592.

Molony, D. (1997). Oftel lobbies MEPs over equal access. *Communications Week International, 195*, November 24, p.9.

Noll, R. (1989). Telecommunications regulation in the 1990s. In P. Newburg (Ed.), *New directions in telecommunications policy* (pp. 11-48). Durham, N.C., and London: Duke University Press.

Noll, R., & Smart, S. (1989). *The political economics of state responses to divestiture and federal deregulation in telecommunications* (Discussion Paper No. 148). Palo Alto, CA: Workshop of Applied Microeconomics, Stanford University.

Public Network Europe. (1997). Judgement day approaches. *Public Network Europe, 7*(11), 32.

Ravaioli, P. (1991). La Communauté Européenne et les Télécommuications: Développements récents en matière de Concurrence [The European Community and telecommunications: Recent developments in competition issues]. *Revue Internationale de Droit Economique, 2*.

Schenker, J. (1993). EC gives in to pressure. *Communications Week International, 102*, April 5, p.1.

Vogelsang, I. (1994). *Federal versus state regulation in U.S. telecommunications* (Monograph No. 134). Bad Honnef, Germany: Wissenschaftliches Institut für Kommunikationsdienste.

8

Operator, Please Give Me Information: The European Union Directive on Data Protection in Telecommunications

Viktor Mayer-Schönberger
University of Vienna, Austria, and Harvard University

The European Union's (EU's) drive for an information society has liberalized a gigantic telecom market, but the EU has not only liberalized the market; it has also moved to strengthen the privacy claims of individuals by drafting a general omnibus Data Protection (DP) Directive and a sector-specific Telecom DP Directive. These directives set forth a comprehensive data protection regime based on stringent processing principles and individual participation rights. Because of the limitations imposed on transferring personal data outside the EU, telecom operators in the United States who operate a point of presence in Europe will in most cases have to comply with these regulations.

INTRODUCTION

On December 15, 1997, the EU passed a "Directive concerning the processing of personal data and the protection of privacy in the telecommunications sector."[1] This directive marks the end of a legislative process lasting more than a decade, with the potential not only to seriously affect the telecommunication sector in Europe, but to have substantial repercussions on U.S. telecommunication companies operating in Europe.

---

1. Directive 97/66/EC of the European Parliament and of the Council of 15 December 1997 concerning the processing of personal data and the protection of privacy in the telecommunications sector. Available: http://www2.echo.lu/legal/en/dataprot/protection.html.

In this chapter I first sketch out the context in which the directive came about—institutional, economical, and legal—and its conceptual foundations. Then I describe the main substantive norms embodied in the directive and how they are related to more general principles discussed earlier. Finally, I briefly consider the important jurisdictional issues involved.

CONTEXT

Over the last two decades the EU, which comprises 15 nations[2] and always intended to be more than just a free trade zone, has moved not only in terms of acronyms from a European Economic Community (EEC)[3] to a European Union[4]. With the Single European Act[5], the EU has broadened and deepened its scope.

The EU is founded on four basic principles, or "freedoms" guaranteed throughout its 15 member states: the free movement of goods, of services, of capital and of labor.[6] To achieve these freedoms, the EU has enacted legislation to both eliminate national restrictions and to harmonize differing laws among member states.

Economical Aspects

The European telecommunication sector in 1996 was 186 billion ECU strong.[7] In 1998 it should reach 196 billion ECU. In the United States, the telecommunication sector accounted for 168 billion ECU in 1995 in a global market of approximately 600 billion ECU. This makes the EU market not only larger than the United States, but also the largest regional market in the world for telecommunication services.[8]

Within Europe, Germany commands a share of 25% of the entire EU market.[9] The German, Italian, French, British, and Spanish telecom sectors taken together account for almost three quarters of the whole European telecom market.[10]

---

2. In alphabetical order: Austria, Belgium, Britain, Denmark, Finland, France, Germany, Greece, Ireland, Italy, Luxembourg, Netherlands, Spain, Sweden, and Portugal.

3. Treaty establishing the European Community, signed March 25 1957, as amended at http://europa.eu.int/abc/obj/treaties/en/entoc05.htm.

4. Title I Article A, Treaty on European Union, signed February 7 1992, http://europa.eu.int/abc/obj/treaties/en/entoc01.htm.

5. Single European Act, OJ L 169 of June 29, 1987, http://europa.eu.int/abc/obj/treaties/en/entr14a.htm.

6. Article 7a of the Treaty Establishing the European Community (as amended), http://europa.eu.int/abc/obj/treaties/en/entr6b.htm.

7. 1 ECU (European currency unit, in many ways the predecessor to the Euro) equaled 1.0935 U.S. dollars as of August 28, 1998.

8. Europe: 31%, United States: 28%, Japan: 14%, Asian Tiger nations 5%, other countries 22%.

9. Germany (1996): 42 billion ECU (Source: ITU).

The European telecommunication sector is not only large, but also has been growing at an annual rate of roughly 8% for the last 5 years. Size and growth taken together with the following institutional reasons have made it an important target for EU policies.

Institutional Aspects

According to EU belief, abolishing customs and harmonizing laws leads to a robust internal market. Such measures will not suffice, however, in sectors, in which state-owned and operated companies have traditionally held a monopoly. These sectors will require special regulations to not only liberalize the markets for new entrants, but to reregulate them so that private sector players may enter without being stifled by unfair competition practices from former monopolists. For the EU, the telecommunication sector is a test case that a transition from a public monopoly to a competitive market of private players can work even when coupled with the conversion from a national to a pan-European regulatory framework.

However, the goal of the EU is even more ambitious. Beginning with the 1994 Bangemann Report, the EU has been aiming at completing what it has termed the *European Information Society* and has called this vision one of the cornerstones of a united Europe.[11] In accordance with this vision, the regulatory framework must not stop at the level of telecommunication regulations, but instead look further at what interactions will take place on these networks.

Through the EU Information Society Project Office,[12] many dozens of legislative initiatives are coordinated. These initiatives taken together form the so-called European Information Society rolling action plan.[13] In the plan, one finds, in addition to the aforementioned core telecommunication regulations, items as varied as taxation, distance selling, digital signatures and encryption, media ownership, and content control on the information and communication infrastructures.

Legal Aspects

In the early 1970s, horrendous scenarios of a total Orwellian surveillance society surfaced amidst gargantuan plans of European (and other) governments to central-

10. France: 26.7 billion ECU; Britain: 23.3 billion ECU; Italy: 20.7 billion ECU; Spain: 11.3 billion ECU (Source: ITU).

11. Compare the so-called Bangemann Report: Europe and the Global Information Society, COM(94)347 final. Available http://www.ispo.cec.be/infosoc/backg/bangeman.html.

12. Available http://www.ispo.cec.be.

13. Draft communication from the European Commission to the Council, the European Parliament, the Economic and Social Committee, and the Committee of the Regions on "Europe at the forefront of the Global Information Society: Rolling Action Plan," COM(96) 607, an updated version is available at http://www.ispo.cec.be/infosoc/legreg/rap2.doc.

ize all personal data about their citizens on central databanks. Ever since then, the desire to protect personal data has played an important role in European legislative developments. As a direct consequence, many European nations enacted national omnibus data protection laws, which in most cases cover both the private and the public sector.[14]

However, in a quarter of a century of passing and amending such national data protection norms, European nations have substantially changed their fundamental approach. Looking at these alterations one can quite easily recognize not only common patterns across national borders,[15] but also specific clusters or generations of similar normative developments. In addition, it is important to note that the U.S. concept of privacy cannot easily be squared with the European concept of data protection. Rather a more complex relationship exists between the two, which can be better assessed by looking at these evolving clusters or generations.[16]

In the first generation, data protection was largely about norms regulating the actual processing of personal data. Second-generation data protection statutes—not the least because of technological and organizational changes in information technology—deemphasized process and stressed individual privacy rights, in the U.S. sense of a negative liberty, of "fencing out" others. The third generation, in the 1980s, arose from the broad realization that in highly interconnected societies, individual privacy cannot be exercised without jeopardizing one's position in society. Thus negative liberties gave way to a concept of data protection based on individual participation rights, particularly empowering individuals to take part in the decision of whether data relating to them may be processed. For this approach the phrase *informational self-determination* has been coined.

However, even the empowerment by participation rights did not bring about the desired outcome of individuals taking the quest for privacy into their own hands. Because of substantial power imbalances and comparatively high costs of litigation, few if any exercised their rights. In a fourth generation, European data protection norms tried to rectify this situation by emphasizing some first-generation procedural safeguards, particularly for specific privacy-endangering sectors, and by amending individual rights to counter implicit power imbalances.

Back in the 1970s the EU decided to harmonize the national data protection and privacy norms. The reason is easy to grasp: According to national laws, personal data could not freely be transferred abroad, outside the scope of the national data

---

14. An excellent collection offers Spiros Simitis et al., *Data Protection in the European Community—The statutory provisions, Nomos* (looseleaf).

15. See Colin Bennett, *Regulating Privacy—Data Protection and Public Policy in Europe and the United States* (1992) and David Flaherty, *Protecting Privacy in Surveillance Societies* (1989).

16. Compare Viktor Mayer-Schönberger, Generational Developments of Data Protection in Europe, in Phil Agre & Marc Rotenberg (Eds.), *Technology and Privacy—The New Landscape* (1997), 219.

protection regimes. Special regulations applied that in effect hindered and limited the free flow of information across Europe. At the same time, the EU was well aware of the need to protect the personal data of its citizens and to ensure a high level of privacy throughout Europe.

Confronted with these needs, the EU drafted an omnibus DP Directive[17] guaranteeing a high level of data protection throughout the EU and, by doing so, creating a harmonized environment within the EU in which personal data, because they are protected throughout, may travel freely across any national border of its member states. The directive, after years of intense debate, was finally adopted on October 24, 1995. It generally applies to all personal data and all stages of its collection, processing, and transfer as long as this is done in a structured (e.g., computer-supported) fashion.[18] The directive had to be implemented by all member states (i.e., transferred into national laws) by October 24, 1998 at the latest.[19]

Furthermore, and in accordance with fourth-generation data protection thinking, the EU recognized that certain economic sectors require additional data protection norms custom tailored to their specific qualities. Consequently, a directive on the protection of personal data in the telecom sector was proposed and after many years of debate adopted in late 1997. Member states will have to transform the directive into national laws by October 24, 1998.[20] A number of states, including Germany and Austria, have already incorporated the directive's text into their national telecommunications regulations.[21]

CONCEPTUAL FRAMEWORK

The principles embodied in the omnibus DP Directive apply to the processing of personal data in all sectors, including telecommunications. Moreover, the sector-specific Telecom DP Directive in Article 2 (1) clearly states that it does not substitute the general DP Directive, but only complements it. It is thus necessary to describe intent, scope, and principles of the general DP Directive to better

---

17. Directive 95/46/EC of the European Parliament and of the Council of 24 October 1995 on the protection of individuals with regard to the processing of personal data and on the free movement of such data, 1995 O.J.(L 281) 31. Available http://www2.echo.lu/legal/en/dataprot/directiv/directiv.html.

18. Art 1 and Art 2 (b) DP Directive.

19. Art 32 (1).

20. Art 15 Telecom DP Directive.

21. Compare Article 2 of the German Federal Act Establishing the General Conditions for Information and Communication Services—Information and Communication Services Act, the so-called Act on the Protection of Personal Data Used in Teleservices (Teleservices Data Protection Act—Teledienstedatenschutzgesetz TDDSG). Available http://www.iid.de/rahmen/iukdgebt.html#a2, § 87 et seq. Austrian Telekomunikationsgesetz, BGBl I/100/1997, http://www.parlinkom.gv.at/pd/pm/XX/I/his/008/I00824_.html.

understand the conceptual framework the Telecom DP Directive fits in.

The general DP Directive envisions a high level of protection of personal data throughout the EU and—because of this guaranteed high level of protection—the completely free flow of personal information between and across member states. Protection is to be achieved by a combination of stringent processing principles and clearly stated individual rights.[22]

Processing Principles

The Directive embodies a two-track approach. The first part ties the legitimate handling of personal data, particularly if structured, through all stages of its use— from collection to processing to storage and transmittal—to specific principles that have to be obeyed by the processor. The most important of these are:

- The purpose limitation principle: Personal data may only be collected for specific, explicit, and legitimate purposes. It must be adequate, relevant, and not excessive in relation to the purposes for which it has been collected. Personal data collected for one purpose may generally not be used for another purpose.[23]
- The data quality principle: Personal data must be accurate and, where necessary, kept up to date.[24]
- The sensitive data principle: The processing of personal data revealing racial or ethnic origin, political opinions, religious or philosophical beliefs, trade union memberships, and health or sex life is generally prohibited.[25] Only under special conditions is the processing permissible.[26]
- The principle of confidentiality and data security: According to the Directive any person who has access to personal data must not process these except on instructions by the controller[27] (basically the one who controls the process

---

22. On the history of the Directive and its largely two-track approach for the protection of personal data within the EU, see, for example, Herbert Burkert, Some Preliminary Comments on the Directive 95/46/EC of the European Parliament and of the Council of 24 October 1995 on the protection of individuals with regard to the processing of personal data and on the free movement of such data, *1996 Cybernews,* No. III. Available http://www.droit.umontreal.ca/pub/cybernews/index_en.html; Peter Blume, An EEC Policy for Data Protection, *1992 Computer/Law Journal 399;* Colin Tapper, New European Directions in Data Protection, *1992 Journal of Law and Information Science 9;* Ulrich U. Würmeling, Harmonization of European Union Privacy Law, *14 Journal of Computer & Information Law 411* (1996).

23. Art 6 (1) b and c DP Directive.

24. Art 6 (1) d.

25. Art 8 (1).

26. Art 8 (2).

27. Art 2 (d).

of collection and processing). The controller is obligated to implement appropriate technical and organizational measures to protect personal data against accidental loss, alteration, unauthorized disclosure, or access.[28]

Individual Rights

In accordance with the concept of fourth-generation data protection norms, the Directive envisions the individual to be by far the best guardian of his or her personal data. Effective data protection, it is the assumption, is best enforced by adamant citizens claiming their rights and legally challenging unfair privacy practices. Consequently, the Directive entrusts the individual with a number of specific privacy rights. In contrast to previous data protection generations, however, the Directive is cognizant of the implicit imbalance between the individual's desire to protect his or her privacy and the overreaching demand of large business and government entities that demand giving up privacy in exchange for contractual services or basic benefits. Thus, the Directive's guarantees have built-in rebalancing mechanisms that allow the individual to exercise his or her rights without the fear or burden of complicated procedures, expensive lawsuits, or unequal negotiations. These rights include, among others:

- The right to collection transparency: When collecting personal data, the individual must be informed about the identity of the controller, the purpose of the processing, and any relevant further information such as recipients of those data or the existence of the right of access to and the right to rectify that data if it is or has become incorrect.[29] Transparency and knowledge are seen as necessary prerequisites for individuals to exercise their privacy rights. Only when one knows that personal data are collected, stored, or processed, can one become involved in the process.
- The right to access data: The controller is obligated to guarantee the right of every individual to access his or her personal data.[30]
- The right to rectify incorrect data: Everyone has the right to have his or her personal data corrected if they are incorrect. [31] This right is an understandable extension of the right to access. What good would it do if you could find out that the data stored about you were incorrect but not rectify that error? It is also an example of the more fundamental idea that the best data protection enforcement is the one effectuated by the individual concerned.

---

28. Art 16 and 17.

29. Art 10.

30. Art 12, exceptions are listed in Art 13; see also the notification duties of the controller in Art 11.

31. Art 14.

- The right to collect damages: In furtherance of this idea, the Directive grants every person who suffered as a result of an unlawful processing the right to receive compensation from the controller for the damage suffered.[32] The compensation is not limited to material damages, but includes immaterial losses as well. In addition, the Directive seems to imply that the individual does not need to prove malice or negligence on the part of the controller to get compensation. Only the controller may, by proving diligence, exculpate himself or herself from liability.

THE EU TELECOM DP DIRECTIVE

Compared with the general DP Directive, the Telecom DP Directive is somewhat different in scope. It applies only to the processing of personal data in connection with the provision of publicly available telecommunication services in public telecommunication networks of the EU.[33] Consequently, the Telecom DP Directive does not pertain to the processing of personal data in nonpublic networks, such as corporate networks. Member states, however, are free to extend protection to these as well.[34] The Telecom DP Directive also extends the scope of the general DP Directive by protecting not only individuals but also legitimate interests of legal persons, insofar as they are subscribers of public telecommunication services or networks.[35]

The Telecom DP Directive is geared toward all telecommunication services, but the language used and the phrases employed in the Directive clearly show that the drafters had almost exclusively thought about point-to-point digital voice telephony, not about a packet-switched Internet.

The Telecom DP Directive employs the same two-tier approach as the omnibus DP Directive by mandating processing principles and laying down a number of individual rights, phrased mostly in terms of participation rights.

Processing Principles

SECURITY

Individuals must be able to trust a telecommunication network and security is an important part of this trust. Thus the Telecom DP Directive obliges telecom ser-

---

32. Art 23.

33. Art 3 Telecom DP Directive.

34. Compare Alexander Dix, Datenschutz und Fernmeldegeheimnis, in Alfred Büllesbach (Ed.), *Datenschutz im Telekommunikationsrecht* (1997), 48.

35. Compare Consideration 6 of the Telecom-DP-Directive; The term *legal person* is to be defined by national law; Alexander Dix, Datenschutz und Fernmeldegeheimnis, in Alfred Büllesbach (Ed.), *Datenschutz im Telekommunikationsrecht* (1997), 48.

vice providers to take "appropriate technical and organizational measures" to safeguard the security of their services and networks. These measures must correlate with the level of risk, but also with the technical state of art and the cost incurred. Here, the final text of the Directive is quite vague. For example, in contrast to earlier drafts, it does not mandate that providers have to offer adequate methods of encryption.[36] Encryption turned out to be too controversial a subject to solve in the process of drafting this Directive.[37]

Particularly noteworthy is the duty of the provider, in case of a breach of security, not only to not cover this incident up, but to inform the individuals concerned about it.[38]

CONFIDENTIALITY

Individuals must be able to communicate confidentially over the network. Thus member states have to provide regulations to ensure the confidentiality of communication networks and telecommunication services. These rules shall prohibit listening to, tapping, storing, or other kinds of interceptions of communications by third parties insofar as they are performed without the consent of the affected users.[39] However, the Directive prescribes neither a constitutional guarantee of confidentiality nor a criminal sanction if confidentiality is violated, instead leaving the implementation to the individual member states.[40]

---

36. See Council of Europe, Recommendation No. R(95) 4 on the protection of personal data in the area of telecommunication services, with particular reference to telephone services.

37. Alexander Dix, Datenschutz und Fernmeldegeheimnis, in Alfred Büllesbach (Ed.), *Datenschutz im Telekommunikationsrecht* (1997), 51; the Organization for Economic and Cooperative Development has not avoided the issue: OECD, Guidelines for cryptography policy, http://www.oecd.org//dsti/sti/it/secur/prod/e-crypto.htm; the EU is currently attempting to formulate a coherent crypto policy in the process of drafting a Directive on Digital Signatures; see Communication from the Commission to the European Parliament, the Council, the Economic and Social Committee and the Committee of the Regions: Proposal for a European Parliament and Council Directive on a common framework for electronic signatures COM(1998) 297. Available http://www.ispo.cec.be/eif/policy/com98297.html; see also the Austrian Draft Digital Signatures Act, which explicitly affirms the free use of cryptography and outlaws key escrow: Viktor Mayer-Schönberger, Michael Pilz, Christian Reiser, & Gabriele Schmölzer, The Austrian Draft Digital Signatures Act, *14 Computer Law & Security Report* (1998) forthcoming.

38. Art 4 (2).

39. Art 5. In an extensive exception, however, Art 5(2) does permit the recording in "the course of lawful business practice for the purpose of providing evidence of a commercial transaction or of any other business communication."

STATUTORY PURPOSE LIMITATION FOR TRANSACTION GENERATED INFORMATION (TGI)[41]

TGI is possibly the most central and publicly most overlooked achievement of modern digital telecommunication networks. With each transaction initiated over the network the subscriber creates a trail of related information that can be traced back to him or her.

The core problem is not the content of TGI itself, but the fact that TGI is so non-transparent to the users.[42] Most of them do not even know that through their transactions they leave behind extensive data trails. To counter this imbalance, the Telecom DP Directive envisions strict limitations for the use and sharing of TGI and desires to empower the consumer to restrict usage of his or her TGI

For example in the area of caller ID, the Directive grants the caller the right to eliminate the presentation of the calling line identification on a per-call basis. In addition, any subscriber of the telecom network service must have the opportunity to block identification on a per-line basis.[43] Similarly, the called subscriber must have the right to block the identification of incoming calls.[44] Furthermore, to allow the called subscriber to block out calls of people not identifying themselves, the subscriber must have the right to refuse any incoming calls of people not identifying themselves.[45]

Similarly, in cases of forwarded calls, the called subscriber must have the possibility to eliminate the presentation of his or her line identification to the caller, thus protecting the privacy of the forwarded-to subscriber.[46]

---

40. The principle of confidentiality is constitutionally guaranteed in a number of European nations. See, for example, Joachim Rieß, Vom Fernmeldegeheimnis zum Telekommunikationsgeheimnis, in Alfred Büllesbach (Ed.), *Datenschutz im Telekommunikationsrecht* (1997), 127; Gabriele Schmölzer & Viktor Mayer-Schönberger, *Das Telekommunikationsgesetz 1997—Ausgewählte rechtliche Probleme,* ÖJZ 1998, 378.

41. The term was coined by Thomas McManus, *Telephone transaction-generated information: Rights and restrictions* (1990); see also Roopali Mukherjee & Rohan Samarajiva, The Customer Web: Transaction Generated Information and Telecommunication, 1993 *Media Information Australia 67,* at 51; Rohan Samarajiva, Interactivity as though Privacy Mattered, in Phil Agre & Marc Rotenberg (Eds.), *Technology and Privacy: The New Landscape* (1997), 277.

42. See, for example, Viktor Mayer-Schönberger, The Internet and Privacy Legislation: Cookies for a Treat?, *1 West Virginia Journal of Law & Technology* (1997). Available http://www.wvjolt.wvu.edu/issue1/articles/mayer/mayer.htm.

43. Art 8 (1)—Presentation and restriction of calling and connected line identification.

44. Art 8 (2).

45. Art 8 (3).

46. Art 8 (4).

All these services must be made available by the telecom network service via a simple means and free of charge. Notably, these provisions do not only apply to calls within the EU. If an individual calls or is called from outside the EU (e.g., the United States), his or her rights to refuse identification and to refuse unidentified calls must remain intact.[47] It is the duty of the telecom network operator to adequately inform the public of these rights available to them.[48] Thus, U.S.-based telecom services need to take the line identification regulations of the Directive into account and adjust their operations to offer such services to EU subscribers, as the Directive does not foresee these subscriber rights to be contractually limited or altered in scope.

STATUTORY PURPOSE LIMITATION FOR BILLING PURPOSES

Generally and in furtherance of the statutory purpose limitation for TGI, all traffic data must be either deleted or made anonymous on the termination of the call that created them.[49] Only for the purpose of subscriber billing and interconnection payments to other telecom operators, the following data[50] may be kept and processed:

- The number or identification of the subscriber station.
- The address of the subscriber and the type of station.
- The total number of units to be charged for the accounting period.
- The called subscriber numbers.
- The type, start time, and duration of the calls made and/or the data volumes transmitted.
- Other related information concerning payments, such as advance payments, installments, disconnection, and reminders.

The keeping of such data is permitted only up to the end of the period during which the bill may lawfully be challenged or payment may be pursued.[51] Furthermore, the operator may use these data for marketing its telecommunication services only if the subscriber has given his or her express consent.[52] A simple "opt-out" possibility will not be sufficient. Subscribers need to expressly "opt-in" for the data to be used in marketing.

For U.S. telecommunication companies doing business in Europe, these limitations on the use of billing and traffic data may require substantial organizational or technical changes in the network to ensure compliance.

---

47. Art 8 (5).
48. Art 8 (6).
49. Art 6 (1).
50. To be found in a separate annex to the Directive.
51. Art 6 (2).
52. Art 6 (3).

Individual Rights

ITEMIZED BILLING

Itemized billing is not common among European telecommunication providers, but the EU views it as desirable to ease checking bills and comparing services.[53] At the same token and furthering the participatory approach of fourth-generation data protection conceptions, the Telecom DP Directive stipulates that individual subscribers have to be given the right to request a nonitemized bill.

To address the privacy issue even with itemized bills, the Directive empowers member states further to "apply national provisions in order to reconcile the rights of subscribers receiving itemized bills with the right to privacy of calling users and called subscribers, for example by ensuring that sufficient alternative modalities for communications or payments are available to such users and subscribers."[54] Such alternative modalities may include the deletion of the last digits of the phone numbers listed in the itemized bill or the encouragement of the development of alternative payment facilities that allow anonymous or strictly private access, like prepaid phone cards.[55]

For the vast majority of U.S. telecommunication operators active in Europe, itemized billing will be familiar. What might turn out to be substantially more tricky is amending the complex billing systems to implement the national variations of nonitemized or semianonymous itemized billing foreseen in the Directive.

CALL FORWARDING

Modern digital telecom networks allow for calls to be automatically forwarded to any given subscriber. To protect any subscriber against the nuisance caused by someone else automatically forwarding calls to him or her, each subscriber must be given the ability to stop automatic call-forwarding via a simple means free of charge.[56]

DIRECTORIES OF SUBSCRIBERS

Individuals shall be able to determine the extent to which their personal data are published in a phone or similar directory.[57] Consequently, the information given

---

53. Common Position Adopted by the Council on 5 June 1997 with a View to Adopting Council Directive of the European Parliament and of the Council on the Application of Open Network Provision (ONP) to Voice Telephony and on Universal Service for Telecommunications in a Competitive Environment. Available http://www.ispo.cec.be/infosoc/telecompolicy/en/Main-en.htm.

54. Art 7 (2).

55. Consideration 16.

56. Art 10.

57. Consideration 19.

in printed or electronic directories is restricted to what is necessary to identify a particular subscriber.[58] In addition, any subscriber has a right to be omitted from such a directory or to restrict information about himself or herself in directories free of charge. If any additional personal data of the subscriber are to be published, the unambiguous consent of the subscriber to do so is required.[59]

UNSOLICITED CALLS

With the advent of digital telecommunication networks and the ubiquity of phones and fax machines, direct marketing has discovered their usefulness in communicating more directly with people who have learned how to get rid of conventional direct ("junk") mail. Automatic calling machines and fax polling stations have made using these lines of communication even more affordable for direct marketing purposes.

The European Union Distant Selling Directive,[60] which applies to business-to-consumer relationships, prohibits in Article 10 the use of voice mail systems and fax machines for direct mail purposes unless the consumer has given his or her prior consent. The Telecom DP Directive extends the protection to business-to-business relations.[61] Automatic calling systems without human intervention (automatic calling machines) or fax machines used for the purposes of direct marketing can only call subscribers who have given their prior consent to get such calls.[62]

Unsolicited calls for purposes of direct marketing employing nonautomatic means are prohibited if either the subscriber has not consented or the subscriber does not wish to receive such calls. The choice between an "opt-in" system based on consent or an "opt-out" one is to be determined by national legislation.[63] Blocking of such unsolicited calls, though, must be free of charge to the subscriber.[64]

---

58. Art 11.

59. Member states may according to Art 11 (3) make such rights available only to natural persons.

60. Directive 97/7/EC of the European Parliament and of the Council of 20 May 1997 on the Protection of Consumers in Respect of Distance Contracts, 1997 OJ (L 144), 24. Available http://www.europa.eu.int/en/comm/dg24/cad/dir1en.html.

61. Giovanni Buttarelli, European Union states reach political agreement on EU telecoms (ISDN) directive, *Privacy Laws & Business Newsletter,* September 1996, 23.

62. Art 12 (1).

63. Giovanni Buttarelli, European Union states reach political agreement on EU telecoms (ISDN) directive, *Privacy Laws & Business Newsletter,* September 1996, 23.

64. Art 12 (2); These restrictions may in accordance with Art 12 (3) be limited by member states to natural persons.

ENFORCEMENT AND JURISDICTION

The data protection regime envisioned by the EU permits the free flow of personal information within the EU not only by harmonizing the national data protection statutes at a high level of protection, but also by severely limiting the transfer of any personal data outside the member countries. Article 25 of the general DP Directive prohibits all transfer of any personal data to a third country, except if an adequate level of data protection is guaranteed in this third country. The adequacy of protection has to be assessed in light of all circumstances, particularly the nature of the data, the purpose of the proposed processing operation, the rules of law in force both general and sectoral in the third country, and the professional rules and security measures that the processor complies with in this third country.[65]

As the United States does not have an omnibus federal data protection statute and rather limited sectoral data protection norms, the burden to show an adequate level of protection rests almost solely on the businesses desiring to transfer to and process personal information in the United States.[66] In essence, private companies will have to show to regulatory bodies in Europe that they comply with and abide to—more or less—EU data protection standards as set forth in the DP Directives.[67]

To somewhat ease this very substantial burden on companies outside the EU, Article 26 of the DP Directive allows personal data to be transferred to a third country without the requirements of Article 25 to be met. These exceptions apply in cases in which

- The individual has given his or her explicit consent to the transfer.
- The transfer is necessary for the performance of a contract between the individual and the "controller" of the data processing (the "controller" needs to control both collecting and processing within the EU and in the third country).

The Telecom DP Directive takes this tough approach one step further by including within its scope all telecommunication services that are publicly available throughout the EU, thus covering, for example, U.S. long-distance operators with just one European point of presence.[68] For other issues of transborder flow of personal data, for instance if telecom billing is centralized in the United States, the provisions of the general DP Directive as detailed earlier apply.[69]

---

65. Art 25 (2).

66. National regulatory bodies are entrusted by the Directive to draw up a list of countries that provide adequate protection (Art 21 (2)). Given the current lack of data protection norms, the United States will not be part of such a list.

67. Art 26 (2).

68. Furthermore line identification (caller ID) suppression has to be guaranteed by the service provider even for calls outside the EU, as long as one subscriber is residing within the EU. See *supra*.

As a consequence, U.S. providers will have to adopt stringent data protection principles in processing personal data or be prepared to demonstrate explicit and unequivocal consent of their customers to the transfer of all such data to the United States.

CONCLUSION

The EU's drive for an information society has liberalized a gigantic telecom market, but the EU has not only liberalized the market; it has also moved to strengthen the privacy claims of individuals by drafting a general omnibus DP Directive and a sector-specific Telecom DP Directive. The regulations set forth in these directives must be well understood and taken seriously, even by businesses abroad that so far have been unaware of such limitations and hurdles. U.S. lawyers might, once again, find themselves digging into EU legislation when aiding their clients at home.

ACKNOWLEDGMENTS

Parts of this chapter are based on an article written with Dr. Gerhard Laga, entitled Tele-Privacy: The European Union Directive on Data Protection in Telecommunications and Its Likely Transatlantic Impact, *Loyola Intellectual Property & High Technology Law Quarterly,* 2(1), (1998)

REFERENCES

Bennett, C. (1992). *Regulating privacy: Data protection and public policy in Europe and the United States.* Ithaca, NY: Cornell University Press.

Blume, P. (1992). An EEC policy for data protection. *Computer/Law Journal, 11,* 399–426.

Burkert, H. (1996). Some preliminary comments on the Directive 95/46/EC of the European Parliament and of the Council of 24 October 1995 on the protection of individuals with regard to the processing of personal data and on the free movement of such data. *Cybernews,* III. Available http://www.droit.umontreal.ca/pub/cybernews/index_en.html

Buttarelli, G. (1996, September). European Union states reach political agreement on EU telecoms (ISDN) directive. *Privacy Laws & Business Newsletter,* p. 23.

---

69. Cf Joachim Jacob, Datenschutzkontrollen über die Grenzen hinweg, in Helmut Bäumler (Ed.), *"Der neue Datenschutz"—Datenschutz in der Informationsgesellschaft von morgen* (1998), 109, 117–118.

Dix, A. (1997). Datenschutz und Fernmeldegeheimnis [Data protection and telephone privacy]. In A. Büllesbach (Ed.), *Datenschutz im Telekommunikationsrecht* (p. 48).

Flaherty, D. (1989). *Protecting privacy in surveillance societies.* Chapel Hill, NC: University of North Carolina Press.

Jacob, J. (1998). Datenschutzkontrollen über die Grenzen hinweg [Control of data protection beyond national borders]. In H. Bäumler (Ed.), *"Der neue Datenschutz"—Datenschutz in der Informationsgesellschaft von morgen* (pp. 109, 117–118).

Mayer-Schönberger, V. (1997a). Generational developments of data protection in Europe. In P. Agre & M. Rotenberg (Eds.), *Technology and privacy: The new landscape* (p. 219).

Mayer-Schönberger, V. (1997b). The Internet and privacy legislation: Cookies for a treat? *West Virginia Journal of Law & Technology, 1.*

Mayer-Schönberger, V., Pilz, M., Reiser, C., & Schmölzer, G. (1998). The Austrian Draft Digital Signatures Act. *Computer Law & Security Report, 14.*

McManus, T. (1990). *Telephone transaction-generated information: Rights and restrictions,* Program on Information Resources Policy, Harvard University, Cambridge.

Mukherjee, R., & Samarajiva, R. (1993). The customer web: Transaction generated information and telecommunication. *Media Information Australia, 67,* 51.

Rieß, J. (1997). Vom Fernmeldegeheimnis zum Telekommunikationsgeheimnis [From telephone privacy to telecommunications privacy]. In A. Büllesbach (Ed.), *Datenschutz im Telekommunikationsrecht* (p. 127).

Samarajiva, R. (1997). Interactivity as though privacy mattered. In P. Agre & M. Rotenberg (Eds.), *Technology and privacy: The new landscape* (p. 277). Cambridge, MA: MIT Press.

Schmölzer, G., & Mayer-Schönberger, V. (1998). Das Telekommunikationsgesetz 1997—Ausgewählte rechtliche Probleme [The Telecommunications Act 1997—selected legal problems]. *Österreichische Juristenzeitschrift 1998,* 378.

Simitis, S., et al. (1992, looseleaf). *Data protection in the European Community—The statutory provisions.* Baden-Baden, Germany: Nomos.

Tapper, C. (1992). New European directions in data protection. *Journal of Law and Information Science, 3,* 9–24.

Würmeling, U. (1996). Harmonization of European Union privacy law. *Journal of Computer & Information Law, 14,* 411.

9

The Architectures of Mandated Access Controls

Lawrence Lessig
Harvard Law School

Paul Resnick
University of Michigan

This chapter proposes an abstract model of mandated access controls for the Internet. The model includes three types of actors: senders, intermediaries and recipients. Control decisions are based on three types of information: the item, the recipient's jurisdiction, and the recipient's type. With the architecture of today's Internet, any party on whom responsibility might be placed has insufficient information to carry out that responsibility. That architecture could be changed to provide senders and intermediaries more information about recipient jurisdiction and type or to provide recipients and intermediaries more information about item types. Although such changes are possible, they would be costly in terms of time, money, and freedom. Moreover, such changes would have side effects of enabling regulation of the Internet by both public and private entities, beyond the scope of any legitimate government interest in controlling access to information.

INTRODUCTION

Speech, it is said,[1] divides into three sorts—(a) speech that everyone has a right to (political speech, speech about public affairs), (b) speech that no one has a right to (obscene speech, child porn); and (c) speech that some have a right to but others do not (in the United States, *Ginsberg* speech, or speech that is "harmful to mi-

---

1. See Lawrence Lessig, *What Things Regulate Speech: CDA 2.0 vs. Filtering,* Jurimetrics Summer 1998, at 10.

nors," to which adults have a right to but children do not). Speech protective regimes, in this view, are those in which the first speech category predominates; speech repressive regimes are those in which the last two speech categories prevail.

This divide is meaningful within a single jurisdiction. It makes less sense when thinking about speech across jurisdictions. For when considering speech across jurisdictions, most controversial speech falls into the third category—speech that is permitted to some in some places, but not to others in other places. What is "political speech" in the United States (Nazi speech) is banned in Germany; what is "obscene" speech in Tennessee is permitted in Holland; what is porn in Japan is child porn in the United States; what is "harmful to minors" in Bavaria is Disney in New York. Every jurisdiction has some speech to which access is controlled,[2] but what that speech is differs from jurisdiction to jurisdiction.

This diversity in the regulation of access creates special problems when we consider speech in cyberspace, for the "architecture" of cyberspace makes selective, or jurisdiction-specific access controls difficult. By architecture, we mean both the Internet's technical protocols (e.g., TCP/IP) and its entrenched structures of governance and social patterns of usage that themselves are not easily changed, at least not without coordinated action by many parties. So understood, the initial architecture of the Internet makes jurisdiction-specific access controls difficult because within this initial architecture, the identity and jurisdiction of the speaker and receiver are not easily tracked either by law enforcement or by the participants in conversation. Neither is the content of speech readily identifiable as it crosses jurisdictional boundaries. As a result, real space laws become difficult to translate into cyberspace. Put another way, the initial architecture of cyberspace in effect places all speech within the first category.

One possible, if unlikely, response to this initial feature of the Internet would have been for governments simply to give up on access control. However, experience suggests that this is unlikely. As the popularity of the Internet has grown, governments have shown an increasing interest in re-establishing mandated access controls over speech. In the United States, this speech is sex-related;[3] in Germany, it is both sex- and Nazi-related;[4] in parts of Asia, it is anything critical of Asian governments.[5] Across the world governments are moving to reregulate access to speech in cyberspace, so as to reestablish local control.

---

2. We reserve the term *censorship* for blanket restrictions on the distribution of speech that apply regardless of the recipient or context. Access control is a broader concept that includes censorship but also restrictions on speech that may depend on the recipient or context.

3. See Telecommunications Act of 1996, Pub. L. No. 104–104, Title V, 110 Stat. 56, 133–43 (1996) (Communications Decency Act).

4. See Kim L. Rappaport, *In the Wake of Reno v. ACLU: The Continued Struggle in Western Constitutional Democracies with Internet Censorship and Freedom of Speech Online*, 13 Am. U. Int'l L. Rev. 765 (1998).

We take as given this passion for reregulation; we consider it a feature of the current political reality surrounding cyberspace. This reality should push us to consider the options that regulators face—not because regulators need to be encouraged toward their regulatory end, but because we should understand the consequences of any particular regulatory strategy. Some strategies are more costly than others; some strike at features of the Internet that are more fundamental than others.

Our aim in this chapter is to offer a way to think about the trade-offs among the various ways in which government-mandated access control could be grafted onto cyberspace. Given that different jurisdictions will want different restrictions, and given that those restrictions would be differentially costly, we provide a map of the different architectures and assignments of responsibility that might effect these restrictions and the trade-offs among these alternatives.

The approach is a type of sensitivity analysis. Regulation, in the view we take of it here, is a function of both law and the architecture of the Internet within which law must function. This architecture itself is subject to both direct and indirect regulation by law. We ask first how access can be controlled given the existing array of legal and architectural constraints. We then consider, second, how changes in the current array might yield a different mix of costs and benefits.

We evaluate the various outcomes of these different legal and architectural choices along four separate dimensions. For any particular mix, we consider first the effectiveness at controlling access; second, the cost to participants, whether sender, or receiver, or intermediary; third, the costs to a system of "free speech"; and fourth, other second-order effects, including in particular how different architectures might enable other regulation, beyond the specific access control that a given change was designed to enable.

For concreteness, we focus on sexually explicit speech. We pick this type because in the U.S. context at least, there are at least two levels of regulation with respect to such speech. Some sexually explicit speech is prohibited generally (obscene speech, child porn), some sexually explicit speech is prohibited only to minors (speech that is "harmful to minors"), and some sexually explicit speech is permitted to everyone. This range of regulations will therefore be illustrative of the more general problem of access control across jurisdictions.

A MODEL OF ACCESS CONTROL: ELEMENTS

In our model of access control, we consider three relevant actors—a sender, a recipient, and an intermediary. The sender is the party who makes available the rel-

---

5. See Geremie R. Barmé & Sang Ye, *The Great Firewall of China*, Wired, June 1997, at 138 and Philip Shenon, *2-Edged Sword: Asian Regimes On the Internet*, N.Y. Times, May 29, 1995, §1 at 1.

evant speech, the recipient is the party who gets access to the relevant speech, and an intermediary is an entity that stands between the sender and the recipient. As these definitions suggest, nothing in our description hangs on whether the sender actually sends material to the recipient. The model is agnostic about the mode with which the recipient gains access.

These actors, we assume, know different things about the speech that is to be regulated. We assume that the sender knows about the contents of the item that is being sent. We assume the recipient has information about who he or she is and where he or she resides. Finally we assume that the intermediary has information neither about the content, nor about who the recipient is or where he or she resides. Obviously, these assumptions are not necessary. A sender might not have knowledge about the speech he or she is making available, and a recipient may not know where or who he or she is. However, we assume a general case.

Given this mix of knowledge, a government effects mandated access control through four separate steps. It first defines which transactions are illegal, where *transaction* means the exchange of speech of a certain kind between two kinds of individuals. Second, it assigns responsibility to one or more actors to effect that restriction. Third, it creates a regime to detect when assigned responsibilities are being violated. Fourth, it sets punishments when these responsibilities are violated. In the balance of this part, we sketch issues relevant to each of these elements of a regulatory regime. In the next part we conduct the sensitivity analysis.

Defining Blocked Exchanges

A regulatory regime first defines a set of illegal transactions, or *blocked exchanges*. The criteria for deciding whether an exchange is blocked include: (a) the type of speech item exchanged, (b) the recipient, and (c) the rules of the recipient's jurisdiction. Within this model, there may be "floor" recipients, and "floor" jurisdictions. In the specific context of sexually explicit speech within U.S. jurisdictions, children are a floor recipient type (anything that is permitted to children is permitted to adults as well), and a Bible Belt small town may be a floor jurisdiction (anything that is not blocked there would be permissible everywhere). The two floors can be combined. Anything that is permitted to children in a floor jurisdiction is permitted to everyone in every jurisdiction.

In the general case, either the sender's or the recipient's jurisdiction may determine that an exchange is blocked. U.S. laws regulating cryptography, for example, restrict a sender's right to send certain encryption-related material to another jurisdiction; French crypto laws regulate a receiver's right to receive such material.[6] For simplicity, however, we focus on blocked exchanges in the recipient jurisdiction alone. This focus is significant in the context of enforcement, since govern-

---

6. See Stewart A. Baker & Paul R. Hurst, The Limits of Trust 130 (1998).

ments can more easily control their own populations than populations in other jurisdictions.

A jurisdiction, in this model of blocked transactions, may specify that a particular transaction is to be blocked in at least two different ways:

1. The jurisdiction might publish criteria defining what is to be blocked, but require a judgment by the parties about how to apply that criteria. The jurisdiction may or may not then hold parties responsible for correctly making such judgments prior to a determination by the regulating jurisdiction.

2. The jurisdiction may classify specific items as acceptable or blocked for particular recipient types. Such classifications would function as a form of preclearance for allowable speech, with the government promising not to prosecute parties for decisions made in good faith based on the preclearance. Alternatively, the jurisdiction could define a list of prohibited speech. In either case, the determinations of acceptability may occur through a judicial or administrative process, or the jurisdiction may delegate its authority to an independent rating service.[7] A jurisdiction might even rely on a computer program to provide an initial classification of the speech at issue and publish that classification as a preclearance, perhaps with a stipulation that the initial classification might be changed in the future after human review.

In the U.S. context, the ordinary procedure follows Case 1. If a jurisdiction follows Case 2, publishing a list of blocked items for a given recipient type, then the list of items must, ordinarily, be judicially specified.[8] It is unclear whether a regime of voluntary preclearance would be permissible. On the surface, it would seem that any step that would reduce the uncertainty surrounding the distribution of speech would be speech enhancing. On the margin, if speakers could be certain that their speech was permissible, they would be more likely to utter it than if they faced the risk that it would be found to be illegal. However, some who have considered the matter believe that if this voluntary regime became, in effect, mandatory, with speech not appearing on a preclearance list in effect then restricted, it would then become constitutionally suspect.[9] The Constitution notwithstanding, we believe that the net effect on speech is unclear: Lower costs could lead to less

---

7. An example would be CyberPatrol's CyberNot list. See CyberPatrol's *Home Page* (visited January 7, 1999): http://www.cyberpatrol.com/

8. See Paris Adult Theatre I v. Slaton, 413 U.S. 49, 55 (1972) (injunction could be used so long as adequate procedures to determine obscenity had been used).

9. See Frederick Schauer, *Fear, Risk and the First Amendment: Unraveling the "Chilling Effect"*, 58 B.U.L. Rev. 685, 725–29 (1978). The closest case is perhaps Bantam Books v. Sullivan, 372 U.S. 58 (1963), where the Court invalidated a "blacklist" Commission. The preclearance idea is not quite a blacklist—the result of the submission would be a promise not to prosecute, not a determination that the material was "obscene." Again, however, we concede that the line is a difficult one to sustain.

chilling of speech (if it is clearer what is prohibited and what is not) but to more control on speech (if it results in greater prosecution of improper speech).

Assignments of Responsibility

The regulator's second step is to define how best to allocate responsibility among actors to assure that access is controlled. In addition to the sender and recipient, it will sometimes be useful to distinguish among intermediaries. Internet access providers (IAPs), such as America Online or AT&T WorldNet, are the intermediaries closest to the senders and recipients. Internet backbone providers, such as MCI WorldCom and Sprint, carry data between IAPs. Responsibility for controlling access could be assigned either exclusively to one actor or jointly to any combination. In this version of our analysis, we consider only exclusive assignments of responsibility for blocking, although we do consider requiring other parties to provide information to the blocking party.

By hypothesis, no party knows enough to determine whether a particular exchange should be blocked.[10] (Again, the sender does not know the recipient; the recipient does not know the content of the item; the intermediary does not know either.) The law must therefore create an incentive for parties to produce sufficient information to determine whether access should be blocked.

The ordinary ways that the law has for creating incentives are either property or liability regimes. Although a property regime is conceivable, we focus here on a liability regime. The law can create an incentive to produce the information necessary to determine whether an exchange should be blocked by allocating liability to an actor for failing properly to block a transaction, or by setting a default rule about whether to block a transaction when there is uncertainty.

We consider two such defaults. Under the first rule, the sender is liable if he or she enters a transaction that is later determined to be illegal without reliable indicators that in fact the transaction was legal. We call this the *prohibited unless permitted* rule, and as we have formulated it to turn on the steps taken to comply with the law, we believe it is distinct from a prior restraint. Under the second rule, the sender is liable only if he or she enters a transaction that is later determined to be illegal in the face of indicators that in fact the transaction was illegal. This is the *permitted unless prohibited* rule, and it is equivalent to a rule punishing a specific intent to violate the law. One modification of this latter rule would hold the sender responsible if the sender should have known that the transaction is illegal. This would comport with a negligence standard, and we consider this change where rel-

---

10. This does not mean that there would not be extreme, and therefore easy, cases. The speaker would certainly know, therefore, for some kinds of speech that it is highly likely to be permitted, or not. Banalities about the weather are fairly safe speech acts anywhere; sadistic child porn is fairly unsafe in most jurisdictions.

evant in the following analysis. In cases of uncertainty, the prohibited unless permitted rule will be overbroad (it will block more speech than the state has a legitimate interest in blocking), and the permitted unless prohibited rule will be ineffective (there will be insufficient incentive to discover the relevant information about what speech should be blocked).

Our focus, therefore, is on changes that might reduce the uncertainty that actors who are responsible for blocking exchanges face. Stated abstractly, these changes will either tag speech or tag people. If speech is tagged, then it is easier for an intermediary or recipient to determine item types; if people are tagged, then it is easier for an intermediary or sender to identify recipient and jurisdiction types. Our aim here is to consider the various consequences of these different alternatives.

Monitoring and Enforcement

The final two regulatory steps are first, devising schemes for monitoring compliance, and second, implementing schemes for enforcing rules against noncompliant actors. In both cases, where the target of regulation sits, relative to the regulating regime, is an important factor in selecting among regulatory regimes. In the case of monitoring, the technology used to effect the access control will significantly alter the costs of monitoring. Some technologies, that is, would be open for an automated and random verification; others would not.

The major issues for enforcement all involve the question of whether the target of enforcement can easily, or cheaply, be reached. We assume there are more receivers than senders, so one might believe targeting senders would be cheaper than targeting receivers. This, however, is complicated if the sender is outside the regulating jurisdiction, making the sender sometimes legally, and if not legally, then often practically, beyond the reach of the regulating jurisdiction. The cost of enforcement against aliens may mean that it is cheaper to enforce a rule against receivers than senders.

Whether there are more receivers or listeners, however, there are certainly fewer intermediaries than either. Intermediaries, as we discuss later, may therefore be the optimal target of regulation, even though they have even less information than either the sender or receiver. Again, the savings of enforcing a rule against them may be greater than the cost of their obtaining the necessary information.

ALLOCATING RESPONSIBILITY

We now consider the consequences of various allocations of responsibility among our three actors, and within each, how changes in existing law and architecture might better achieve the aim of access control, with fewer free speech costs, or more access control effectiveness.

Sender Responsible for Blocking Access

Our first rule would make the sender responsible for controlling access. To comply, the sender must determine both the law of the jurisdiction of the recipient, and depending on that law, the character of the recipient. Under existing state law, for example, it is often prohibited to distribute material that is "harmful to minors" to minors. Under our first rule, then, the sender must determine whether the recipient comes from such a jurisdiction and if so, whether he or she is a minor.

Under the present architecture, both determinations are costly. There is no simple way to identify the jurisdiction within which the recipient resides and no cheap way to be certain of characteristics of the individual. The rule would therefore be quite costly to a speaker—unconstitutionally costly is the suggestion of *Reno v. ACLU*—although the costs would be different under each default rule.

Under the prohibited unless permitted rule, the cost is to "free speech" interests. The burden of determining eligibility is likely to present a significant chill on the speaker's speech. The sender would have to take steps outside of the architecture of the Internet to determine where a recipient is—by verifying an address, for example, or using an area code on a telephone number as a proxy. The sender would need to rely on proxies from credentials (such as a credit card) to guess whether the individual is a proper age or not.

If, on the other hand, the rule is permitted unless prohibited, the cost is the effectiveness of the regulation. Under this rule, the existing architecture would make any access control ineffective. Whereas in real space, certain facts about an individual are unavoidably self-authenticating (a 10-year-old boy does not look much like a 20-year-old man), in cyberspace, such facts are not self-authenticating. To determine either the jurisdiction or the age of the recipient requires affirmative steps by the sender. If no obligation to take such steps exists or if no requirement exists to block unless such steps are taken, then the rule will not effect the intended access control.

The existing architecture therefore creates a great burden for the sender if the default is prohibited unless permitted, and it defeats access control if the default is permitted unless prohibited.

SENSITIVITY

Some of the burden on the sender created by this rule could be reduced by certain architectural and legal changes. In this section we describe four, and consider the potential costs and benefits of each.

The first two changes involve less expensive ways to identify facts about the recipient. The two facts unknown by the sender are the jurisdiction and characteristics of the recipient (e.g., he or she is over 18). The changes described here would facilitate the sender knowing both facts at a relatively low cost.

The first technique relies on digital certificates. In the standard model of certificates, *certificates* identify who someone is. They are digital objects cryptographically signed by a certificate authority (CA). The dominant use of such certificates today is to certify the identity of the holder. This is the model, for example, of the Verisign Digital ID, which Verisign describes as a "driver's license for the Internet."

However, there is no reason that the same technology could not be used to certify facts about the holder—or, more generally, to certify any assertion made by the signer. In our case, a signing CA could then certify that X is from Massachusetts, and that X is over the age of 18, without identifying who X is.[11] Senders would then examine these certificates before granting access to regulable speech. Access would then be granted without a cumbersome system of passwords or IDs.

We can call this a *credentialing* solution.[12] It requires that the sender make certain judgments about the speech at stake, but it allows the sender to rely on representations about the jurisdiction and the recipient that are necessary to determine whether an exchange is or is not blocked.

Under a prohibited unless permitted regime, access would be blocked except to those who could show that they carry the proper credentials. In the case of harmful to minors speech, the credential would be an adult ID indicating that the recipient is over 18. Recipients interested in receiving restricted materials will have an incentive to show such credentials. All else being equal, certificates would lower the cost of such a showing and therefore reduce the burden, and hence chill, of the access control regime. Moreover, the burden on individuals under such a regime would be lower than under a regime where they must show a credit card or other form of identification. The cost of a certificate should be less than the cost of a card and the possibilities for anonymity should be greater.

No legal mandate on recipients will be needed to encourage showing age or jurisdiction certificates under a prohibited unless permitted regime, but sanctions would be needed to reduce the use of fraudulent certificates. If, for example, it were easy to obtain an anonymous adult ID certificate, one might imagine a black market emerging with children acquiring certificates from adult intermediaries. One way to limit the transferability of anonymous certificates would be to include an Internet protocol (IP) address in the certificate so that it could only be used with a single computer, or for the duration of a single dial-up connection if an access provider assigns different addresses for each dial-up session. Another technique for limiting transfers would be to make the certificates traceable so that if abuse is

---

11. David Chaum was an early proponent of such characteristics certificates rather than identity certificates. See David Chaum, *Security Without Identification: Transaction Systems To Make Big Brother Obsolete*, 28 Comm. of the ACM 1030 (1985).

12. Note that even though the technology for this solution is already in place, we refer to it as a possible architectural change because a widespread change in social practices would be necessary for the technology to be used in this way.

detected, the identity of the original acquirer could be revealed and that person could be punished.

Alternatively, widespread use of digital certificates could also improve the effectiveness of a permitted unless prohibited regime by providing senders with enough information correctly to block exchanges that would otherwise have been permitted by default. For harmful to minors speech, the credential would be a child ID indicating the holder was under 18. In this case, however, recipients (if children) would have no natural incentive to provide a child ID, since such credentials can only cause otherwise permitted exchanges to be blocked.[13] It follows that to achieve widespread use of jurisdiction and recipient type credentials under a permitted unless prohibited regime, it would be necessary to impose additional legal rules that require recipients, in certain situations, to provide relevant credentials.

To minimize the burden of this rule, the rule could require that the recipient provide the certificate only if the server asks, and the server asks only if the material is illegal in at least one jurisdiction. This regime would still burden somewhat those recipients living in jurisdictions where the speech was wholly legal; its viability would rest then on the significance of that burden.[14] Alternatively, the rule could require that intermediaries provide or assure that users have valid certificates. In this case, the appropriate intermediaries would be the IAPs who serve recipients. If the state requires such intermediaries to assure the supply of certificates then the cost of monitoring and compliance might be lower than if the same role were being performed by the state. The intermediary's advantage is not over the primary conduct—certainly receivers are in a better position to certify than intermediaries—but in assuring that the primary conduct is properly regulated.

A second architectural change to help the sender identify the jurisdiction into which speech was to be sent would be an IP map—a table that would give a rough approximation of the location of the recipient's computer.[15] No doubt the map could not be perfect, and senders or recipients could use proxies to escape the con-

---

13. Parents or guardians, however, might have a sufficient incentive to configure software to always pass a child ID certificate to servers when children are accessing the Internet.

14. Another possibility would be for the server to send a request of the form "if you are in jurisdiction X or Y and you are under 18, please provide a child ID," which would further reduce the burden of the system.

15. Currently, the InterNIC maintains a database of which organization each IP address was allocated to. This database is public and a copy of it may be requested from any computer on the Internet. Unfortunately, some entries in the database are incomplete or out of date, and they do not necessarily identify the location of computers using the IP addresses. It has been suggested, however, that this database be used as a starting point for developing an IP to jurisdiction mapping. See Philip McCrea, Bob Smart, & Mark Andrews, *Blocking Content on the Internet: A Technical Perspective*, Appendix 5 (visited August 22, 1998). Available http://www.noie.gov.au/reports/blocking/index.html.

sequences of the map, but in the main, the map might sufficiently segregate restrictive jurisdictions from those that are nonrestrictive.[16]

An IP map would provide benefits over a certificate system. Under the prohibited unless permitted regime, an IP map may burden speech even less than the certificate regime, since the cost to the recipient of this form of identification is zero, and the processing costs to the server would be lower than processing a certificate. The permitted unless prohibited regime becomes more effective as well, since now the sender has an assured way of knowing the jurisdiction into which the material is being sent, but not information about the recipient's age or other characteristics.

However, there are important social costs associated with this IP-to-geography mapping that flow from its generality. Because jurisdiction identification would be determinable with any IP transaction, the regime would effect jurisdiction identification independent of the kind of speech being accessed. This raises obvious privacy concerns, which might be mitigated by structures that would limit the use of the list for specific purposes. But for obvious reasons, it would be difficult to limit the use of this information.

We presume that the sender knows about its speech, but it may not understand the classification scheme of every legal jurisdiction. The final two architectural changes would aid senders in classifying their speech according to the categories of various jurisdictions. The first is an automated preclearance technology: A computer program could quickly and cheaply classify speech for particular jurisdictions, although not with perfect accuracy. The second would be a thesaurus that relates the categories of different jurisdictions. Thus, if the sender is able to classify an item according to one jurisdiction's categories, it could infer the classification in some other jurisdictions. For example, it may be that anything classified as child pornography in Jurisdiction A would be classified as obscene in Jurisdiction B, although the converse inference might not hold. The thesaurus functions as a more complex version of the base jurisdiction model that we described earlier.

Recipient Responsible for Not Taking Access

Our second rule would make the recipient responsible for illegal transactions—that is, targeting the buyer rather than the seller. This rule has some advantages over the sender-responsible rule—the recipient, for example, may be in a better position to know about the law of its jurisdiction and about its own recipient type.

---

16. We note that already companies such as Microsoft are using IP addresses to assure themselves that the user is within the United States, so that Microsoft does not become an "exporter" of high-grade encryption technology. See *Internet Explorer Products Download: Microsoft Strong Encryption Products (US and Canada Only)* (visited August 23, 1998). Available http://mssecure.www.conxion.com/cgi-bin/ieitar.pl.

There are obvious disadvantages as well. The recipient is in a worse position, relative to the sender, to know about the kind of information that the sender is making available. A sender may find it burdensome to classify its speech according to a particular jurisdiction's categories, but at least the sender begins with knowledge about the content of the speech at issue.[17] The receiver does not. This means that a recipient cannot determine the legality of an exchange until after the exchange has occurred. Thus, under a prohibited unless permitted rule, the receiver risks liability[18] in the very act of determining whether a particular exchange complies with the law. If the rule is permitted unless prohibited, then restrictions are likely to be completely ineffective.

A second problem with placing liability on the receiver is simply the costs of classification. The rule shifts the cost of classification to the user rather than the provider so that rather than one person classifying, many are forced to classify.

Finally, putting the responsibility on the receiver may increase the costs of enforcement. Receivers are ordinarily individuals, and a regime that regulates only individuals would either be costly to monitor or dangerous because of selective enforcement. Of course these burdens are contingent, and it could be that they in fact indicate something very different. It may be that given the distribution of recipients and senders, and the likelihood that more recipients will be within the control of the regulating jurisdiction than senders, punishing recipients will be cheaper than punishing senders. Although there might be more recipients than senders, their compliance could be checked randomly; assuming the probability of detection were high enough, their compliance might be easier to secure.

SENSITIVITY

A recipient-responsible rule could be made less costly if there were cheaper ways to identify item types. An obvious solution here is a kind of labeling or rating of items. Two sorts of labeling are possible. One we have already described: prescreening. The other solution is to rely on senders to label their own materials. The labels might directly indicate whether the item is permitted or prohibited to recipients of various ages in particular jurisdictions, or it could describe the item on some dimensions such as sex or hate that were sufficient to infer whether it should be blocked. This solution simply inverts the certificate solution—since here it is

---

17. It would be different, of course, if the sender were considered as a bookstore, without knowledge, or any simple way to get knowledge, about the content of its books. See Cubby v. Compuserve, 776 F. Supp. 135 (S.D.N.Y. 1991). We would consider such a "sender" to be an intermediary in our analysis.

18. This depends on the level of knowledge required for someone to be guilty under such a provision. If the statute were criminal, the knowledge requirement would be quite strong, so inadvertent liability would not be possible. But for a lesser prohibition, the knowledge requirement may be less.

the sender that is offering a "certificate" and the receiver who is relying, whereas in the earlier case, it was the recipient providing the certificate, and the sender who was relying. The analysis is also analogous.

Under a prohibited unless permitted regime, the labels would convey information that the speech was permitted (e.g., no sex or hate speech). Recipients would be given immunity if they in good faith rely on a sender's labels to determine that access is permitted. Senders would have a natural incentive to provide labels, since they would allow more recipients to receive the speech, although penalties for inaccurate labels might be needed to prevent widespread mislabeling. There would of course be a transition period, during which only a small percentage of materials would carry self-rating labels, rendering most of the Internet blocked under a strict prohibited unless permitted rule.

Under a permitted unless prohibited regime, the labels would indicate that access to an item was prohibited (to some groups in some jurisdictions). There would be little incentive to label by the sender, since that could only reduce legal access. To bolster the effectiveness of this regime, a government might require senders to provide labels. This might raise a constitutional question in the United States if labels were considered compelled speech. Some have argued that they would not be,[19] but this is a question nonetheless. To reduce the cost to senders of labeling, a government might subsidize third-party rating or itself produce suggested ratings. Its rating could not be definitive[20] in such a system, but might be an aid to senders in self-labeling.

The burden of labels might be minimized by simply requiring labels only where speech is potentially regulable (comparable to requiring that people up to the age of 25 carry IDs to purchase cigarettes, even though the prohibition reaches only those 18 and under). Even here, however, the requirement raises difficult questions, since it is requiring speech by the sender in the form of a label even if the underlying speech is clearly legal in the jurisdiction into which it is being sent. Thus the most restrictive jurisdiction is determining whether the speaker must label.

As with recipient certificates, the responsibility for assuring a supply of sender labels might be assigned to intermediaries, in this case to the sender's IAP. There is one important asymmetry, however. Whereas age and jurisdiction are objective properties that one might reasonably expect an access provider to verify, correct assignment of rating labels to items will involve subjective judgements. One intermediate form of responsibility might be to require an access provider to assure the

---

19. See R. Polk Wagner, *Filters and the First Amendment* (visited August 22, 1998). Available http://www.pobox.com/~polk/filters.pdf.

20. In no case could the government's own ratings be determinative of whether speech were delivered or not, absent a judicial finding. See Rowan v. U.S. Post Office, 397 U.S. 728 (1970).

availability of some sender self-label, but make only the sender and not the IAP responsible for any inaccuracies in the label.

We also note an interesting constitutional asymmetry between requirements of senders providing labels and recipients providing certificates. It seems clear that there could be no U.S. law that required receivers not to receive any speech unless it were plain that the speech was legal (i.e., a rule that would punish the receiver if he or she received speech without a clear indication that it was not illegal). Even with respect to obscenity, that restriction would be overbroad, or fail a minimal *mens rea* analysis. It is at least arguable, however, that a law that required senders to check every digital ID before sending material would not be constitutionally invalid. Analytically these two regimes are quite similar: In both, the transaction is conditioned on verification of its legality. However, the burden on the sender is likely less constitutionally troubling than the burden on the receiver.

Beyond the constitutional issues, there is a practical enforcement problem with mandating that senders provide labels. Just as it may be difficult to enforce blocking requirements across jurisdictional boundaries, it may be difficult for authorities in one jurisdiction to enforce a labeling requirement in another.

Intermediary Responsible for Blocking

We have assumed that the intermediary has information neither about the recipient nor about the item the sender would send. It might therefore seem odd to consider the intermediary as a possibly responsible actor.

However, this intuition is misleading. Intermediaries are a cheap target of regulation. There are fewer of them than receivers or senders, and they are typically more stable or harder to move. Just as it is easier for the government to regulate telephone companies than it is to regulate telephone users, it would be easier for the government to set requirements on intermediaries that intermediaries could then enforce on their customers. More importantly, because intermediaries have an interest in reducing the cost of compliance, regulating intermediaries is more likely to get innovation in the methods of compliance.

In addition to a lack of information, intermediaries may have limited capabilities for implementing blocks. Blocking can either be implemented at the application layer (e.g., Web page requests) or at the network layer (i.e., individual packets). Network layer blocks are of necessity much cruder: Only the sender's and receiver's IP addresses and the port number (a rough indicator of whether the connection is being used for a Web transfer, e-mail, or something else) are available. Thus, a network layer block can either block all Web requests to a particular IP address, or none of them.[21]

---

21. For a more complete description of application layer and network layer blocking, see McCrea, *supra* note 15.

We consider two types of intermediaries. One type is an IAP or service provider or an employer or a school (for simplicity, we refer generically to any of these as an IAP). It is reasonable to assume that an end user and his or her IAP lie in the same jurisdiction.[22] Many but not all IAPs run proxy servers (and other application layer gateways) that intercept some kinds of Internet traffic. Most commonly, a Web proxy at an IAP will keep copies in a cache of frequently accessed Web pages; when a customer requests a cached page, the proxy sends it to the customer, without fetching it again from the sender's Web server. Proxy servers permit application layer blocking: requests for certain universal resource locators (URLs) can be blocked. Moreover, an IAP may configure a firewall that forces all requests to use the proxy server. This is done most frequently to enhance corporate security, by restricting the Internet traffic entering and leaving a corporation to only that which passes through proxies. In those cases where an IAP does not employ proxy servers, however, only cruder network layer blocking is possible.

The second type of intermediary is a backbone provider, which carries data across jurisdictional boundaries. In practice, the IAP may also run backbone services, but the services are conceptually distinct because they have different technical filtering capabilities. Consider the cross-jurisdiction transit point, the place in the backbone provider's network where data cross a jurisdictional boundary. Such transit points do not normally employ proxy servers or other application layer gateways. Thus, only the cruder network layer blocking is possible at cross-jurisdiction transit points, given the current Internet architecture.

One final difficulty with blocking by intermediaries is that recipients may find ways to bypass the blocks, especially if the senders cooperate. For example, the same prohibited document may be available from several different URLs, so that a recipient can access one even if the others are blocked. A technique known as *tunneling,* where the contents of one packet are wrapped inside another packet, may bypass a network layer block.[23]

SENSITIVITY

A combination of the architectural changes discussed in previous sections could provide intermediaries with enough information to decide which exchanges to block. That is, information about item types could come either from senders' labels or from preclearance lists provided by jurisdictions. Information about recipient type could come from certificates and information about recipient jurisdiction could come either from certificates or from a database lookup on the IP address.

---

22. It would be possible, although expensive, to make an international phone call to access an IAP in another jurisdiction.

23. McCrea et al. detail these and other ways that senders and recipients might bypass intermediaries' blocks. See McCrea, *supra* note 15.

One potential change in the architecture to facilitate the implementation of blocking would be to require an application layer gateway at IAPs or cross-jurisdiction transit points and require that all customer traffic use these gateways (perhaps enforced via a firewall). This would have high costs for Internet flexibility and operation. First, it would be computationally expensive to assemble all packets into messages at cross-jurisdictional transit points, especially for traffic where there is no counteracting performance gain from caching. Second, messages may be encrypted for privacy or security purposes (e.g., in connections that use SSL, the Secure Sockets Layer) so that even at the application layer only crude blocks based on sender and receiver address are possible. Third, innovations that introduce new applications would be stifled, since the application layer gateways would not initially know about the new applications and hence would block them.[24] The Internet's current architecture has enabled experimentation and rapid deployment of new applications (examples of applications that blossomed in part as a result of this flexibility include the World Wide Web, ICQ ("I seek you"), and push services). One final cost might come in the form of reliability. It is relatively easy for a service provider to provide multiple routers so that network layer service is not interrupted if a router is temporarily disabled. It may be more costly to arrange for continued service if an application layer gateway is temporarily disabled.

SECONDARY CONSEQUENCES

Our focus so far has been on how to effect mandated access controls. We have considered three different techniques—tagging the sender, tagging the recipient, or regulating the intermediary to help effect either of the two taggings.

The aim in this section is to consider side effects of each strategy. All three strategies envision a general infrastructure that can be used for purposes beyond those initially intended. These we consider side effects of the initial regulatory objective, and they should be accounted for in selecting among strategies for regulating access.

IDs and Regulability

To effect sender or intermediary control, we envisioned the development of identity certificates designed to facilitate the credentialing of certain facts about a recipient—how old that person is, where he or she is from, and so on. We also

---

24. Many corporate firewalls do prevent employees from using experimental applications that the corporate proxy or gateway is not configured to handle. See William R. Cheswick & Steven M. Bellovin, *Firewalls and Internet Security: Repelling the Wily Hacker* 75 (1994).

proposed the development of a database that maps IP addresses to jurisdictions.

However, it should be clear that if these architectures were enabled for this speech-regulating purpose, the certificates and databases would have an application well beyond this purpose alone. These architectures, that is, might facilitate other jurisdiction-based regulation or access control imposed by senders beyond the narrow purposes that motivated the initial change. We might, that is, make the Internet safe for children, but as a consequence make it a fundamentally regulable space.

Certificates or IP databases would facilitate a more general structure of jurisdiction-based control, including taxation and privacy regulations. The reason is fairly straightforward: Local jurisdictions at least have the legal authority to regulate their own citizens, both while the citizens are at home and while they are away. A certificate-rich Internet would facilitate the identification, then, of who could be regulated by whom, or what standards could be imposed on whom. This, in turn, could facilitate a more general regulation of behavior in cyberspace.

We might imagine the model to look something like this: States would enter a compact, whereby they, as a home jurisdiction, agree to require senders or intermediaries within their own jurisdiction and to respect the rules of other jurisdictions in exchange for senders or intermediaries in other jurisdictions doing the same for the home jurisdiction. These rules would specify the restrictions imposed on citizens from a given jurisdiction and the range of citizens for whom the restriction applies. For example, a jurisdiction might specify that citizens from it may not engage in Internet gambling; the jurisdiction within which a gambling server sits, then, would require the server to check for a person's citizenship and condition access based on whether they held the proper credential. Presumably the jurisdiction would do this only if there were restrictions that it wanted imposed in other places and that it needed other jurisdictions to respect.

If a jurisdiction database or a credential-rich Internet were in place, we might then expect voluntary uses of that infrastructure to proliferate. Some voluntarily imposed restrictions might seem reasonable. For example, recording companies might refuse access to their websites from countries where pirated copies of intellectual property were rampant. Other voluntary uses might not have such sanguine effects. For example, some Serbs and Croats might refuse to allow each other access to their Web pages. In both cases, a form of discrimination is being enabled by the certificate infrastructure. The discriminations are then a consequence of this certificate infrastructure.

Labels and Improper Control

The alternative solution that we have identified for effecting mandated access control works to label content so as to facilitate filtering by recipients or intermediaries. The labels might be provided by senders or by governments in the form of preclearance lists. An inexpensive and widely used labeling infrastructure would have its own secondary impacts, including both the possibility of more widespread speech regulation and voluntary individual or collective uses of labels for blocking beyond the state's legitimate interest.

First, if available speech labels describe categories beyond those that a jurisdiction would normally regulate, the mere availability may tempt regulation within these new categories. Thus, the widespread use of a general labeling infrastructure may start governments on a slippery slope toward regulating all sorts of speech, even if the initial impetus for labeling is limited to only a few kinds of speech.[25]

Second, labels might be used for voluntary access controls as well as mandated access controls; that is, recipients or intermediaries might choose to block more exchanges than governments require. Parents in the United States, for example, may choose to block young children's access to hate speech or speech about sex education, even though such speech is legal for children in the United States. Alternately, a search engine may provide a filtered search service that, when queried for *toys,* returns links to pages describing children's toys rather than sex toys, without necessarily reporting that certain sites have been blocked.[26]

The availability of voluntary access controls by parents and teachers is widely viewed as socially beneficial because it gives control to people who can tailor restrictions to individual and local needs. In a world of perfect transparency and competition, such control imposed by IAPs or search engines may be unproblematic as well.

In practice, however, there are a number of reasons these access controls might be less than ideal:

- First, consumers may have a hard time determining which blocks are in their own best interest, as the criteria for selection may not be transparent or readily understandable.[27]
- Second, even if the criteria were transparent, the present architecture would still allow filtering "upstream" (e.g., by a search engine) without the consumer knowing (thus a nontransparency not about the rating, but about who is effecting the filter).[28]

---

25. Obviously, the most significant concern here would be jurisdictions outside of the United States, or outside of places where a strong free speech right exists. The norms that the United States sets for the Internet, however, would certainly spill over into those places, and our view is that this spillover should be reckoned in any regulatory regime.

26. For a demonstration of Alta Vista's Family Filter, using ratings from SurfWatch, click on the AV Family Filter link at http://www.altavista.com/. For a discussion of the implications, see Jonathan Weinberg, *Rating the Net,* 19 Hastings Comm. & Ent. L. J. 453 n.108 (1997).

27. See Rikki McGinty, *Safety Online: Will It Impede Free Speech?* Media Daily, December 5, 1997.

28. See Weinberg, *supra* note 26, at n. 108.

- Third, individuals may face a social dilemma about whether to adopt filters. Individuals may themselves prefer to have filtered content (to perfect their own choice), but not want society to have filtered content (to preserve social diversity).[29] If everyone can easily satisfy his or her individual preference for filtering, the collective preference for social diversity may be ignored.
- Finally, if IAPs bundle filters with service, then the choice among filters might be less robust than ideal. Put another way, in practice, the competition among filters may not be sufficiently diverse. This could yield very broad filters, that if common, could create secondary impacts on the variety of speech available on the Internet—since senders may tailor their speech to what will pass the filters.[30]

These secondary effects—a slippery slope of regulation and potentially chilling voluntary uses of labels—have led one of the authors previously to describe the Platform for Internet Content Selection (PICS), which provides the technical infrastructure for labeling, as "the devil."[31] The other author (one of the developers of PICS) believes that the net impact of a widespread labeling infrastructure would be positive because of the many positive voluntary uses.[32]

CONCLUSION

This chapter has proposed an abstract model of mandated access controls. The model includes three types of actors: senders, intermediaries, and recipients. Control decisions are based on three types of information: the item, the recipient's jurisdiction, and the recipient's type.

With the architecture of today's Internet, senders are ignorant of the recipient's jurisdiction and type, recipients are ignorant of an item's type, and intermediaries are ignorant of both. It is easy to see, then, why, with today's Internet architecture, governments are having a hard time mandating access controls. Any party on whom responsibility might be placed has insufficient information to carry out that responsibility.

Although the Internet's architecture is relatively entrenched, it is not absolutely immutable. Our abstract model suggests the types of changes that could enhance regulability. Senders could be given more information about recipient jurisdiction and type either through recipients providing certificates or through a database

---

29. See Cass R. Sunstein, Democracy and the Problem of Free Speech (1995).

30. See Weinberg, *supra* note 26, at 477.

31. See Lawrence Lessig, *Tyranny in the Infrastructure*, Wired, July, 1997, at 96.

32. See Paul Resnick, *Filtering Information on the Internet*, Scientific American 62 (March 1997) and *PICS, Censorship, & Intellectual Freedom FAQ*, (Paul Resnick, ed.) (last modified January 26, 1998). Available http://www.w3.org/PICS/PICS-FAQ-980126.html.

mapping IP addresses to jurisdictions. Recipients could be given more information about item types either through senders providing labels or through government preclearance lists of permitted or prohibited items.

Table 9.1 summarizes this sensitivity analysis. Because the two interventions are analogous, the analyses of their costs and effectiveness are analogous as well. In either case, there will be a natural incentive to provide information if the default action of the responsible party is to block access unless the information is provided (a prohibited unless permitted regime). Otherwise, there will be no natural incentive, and the government will have to require the provision of that information.

TABLE 9.1
Categorization of Results

	Sender	Intermediary	Recipient
Missing information	• Jurisdiction • Recipient type	• Jurisdiction • Recipient type • Content of item	• Content of item
Possible architectural and legal changes	• IP to geography mapping, jurisdiction certificates • Recipient type certificates • Preclearance, thesauri	As for sender and recipient, plus: • Responsibility to assure sender/recipient compliance • Use of proxies and application gateways	• Preclearance • Sender's self-rating • Third-party rating
Consequences	Enables general regulability of behavior on the Net based on recipient type and jurisdiction	Enables private parties (IAPs and ISPs) to regulate behavior on the Net	Enables greater control of speech content on the Net beyond that initially required by governments
Notes	Enforcement problems significant, if sender outside the jurisdiction	Enforcement is easier, as IAPs are not mobile, there are few players, and they have commercial assets	Enforcement problem: number of recipients leads to selective enforcement, although a greater portion of the regulable public is within a given jurisdiction

The secondary effects of these two infrastructures are also analogous, but quite different. The by-product of a certificate regime is a general ability to regulate based on jurisdiction and recipient characteristics, even for issues beyond content control, such as taxation and privacy. Such a regime also enables senders voluntarily to exclude recipients based on jurisdiction or type, a facility that might be used for negative as well as positive purposes. The by-product of a widely used labeling infrastructure is a general ability to regulate based on item characteristics, even characteristics that governments have no legitimate reason to regulate. Such a regime also enables intermediaries and recipients voluntarily to exclude some item types, a facility that may empower parents and teachers but may also be overused if it is poorly understood or difficult to configure.

If intermediaries are to be responsible for blocking, they will need both types of information. In addition, architectural changes will be necessary to enable application layer blocking of individual items rather than cruder network layer blocking of all traffic from or to an IP address. A requirement of application layer blocking, however, introduces significant costs in terms of openness to innovation and vulnerability to hardware and software failures. Intermediaries, then, are the most costly place to impose responsibility. On the other hand, they are the most easily regulated because there are fewer of them, they are more stable, they have assets, and their governing jurisdictions are clear.

Our sensitivity analysis does suggest consequences that might not have been readily seen, but our ultimate conclusion is one others have reached as well. It will be difficult for governments to mandate access controls for the Internet. Given today's architecture, any such mandates would of necessity be draconian or ineffective. Changes to the technical infrastructure or social practices could enhance regulability, although such changes would entail direct costs and create secondary by-products with debatable value. Given that the costs of any such architectural change would be significant, it is important for governments to answer the fundamental question of how important such changes are: Perhaps a lessening of governments' traditional power to control the distribution of harmful information would be preferable.

ACKNOWLEDGMENTS

Thanks to Lorrie Cranor for initially suggesting the symmetry between tagging speech and tagging people. Alexander Macgillivray provided valuable research assistance.

10

Beyond Infrastructure:
A Critical Assessment of GII Initiatives

Heather E. Hudson
University of San Francisco

At the International Telecommunication Union (ITU) Development and Plenipotentiary Conferences in 1994, U.S. Vice President Al Gore proposed extending the concept of information infrastructure (II) from national to global. The goal of building a Global Information Infrastructure (GII) was soon embraced by international organizations such as the ITU, G7, European Union (EU), and Asia-Pacific Economic Cooperation (APEC), which have also pledged to support trials and pilot projects using new technologies, primarily involving access to the Internet.

Yet GII is typically very vaguely defined, and initiatives to promote it appear to be generally technology-driven without involvement of users or recognition of the constraints imposed by current policies. In these respects, such initiatives appear similar to earlier technology-driven strategies to promote new technologies such as educational television and satellite systems.

THE CALL FOR GII

The phenomenal growth and visibility of the Internet as an information resource, communications tool, and electronic marketplace have focused attention on the need for investment in II to bring the Internet and other forms of electronic communications within reach of people around the world. In 1993, the newly elected Clinton administration called for investment in a National Information Infrastructure (NII) to bring the benefits of advanced services, and particularly the Internet, to Americans not only in the workplace, but in schools, libraries, health care centers, and individual households.

Vice President Al Gore elaborated on this theme at the ITU's Development Conference in Buenos Aires in 1994, calling for a GII to extend access to these new technologies and services to people throughout the developing world. Policymakers in various forums such as the EU, the G7,[1] and APEC[2] have taken up the call, with support for pilot projects and strategies to accelerate investment in telecommunications networks.

THE FIRST STEP: CREATING AWARENESS

To encourage policymakers to support information-based development strategies, it is important to make them aware of the potential of new technologies. Al Gore caught the imagination of traditional telecommunications administrators through his ITU speeches; building on this introduction, the ITU's Telecom 95 Forum included sessions on the Internet for the first time. However, it was also important to reach national decision makers and not just the technocrats.

A major event aimed at industrialized country leaders was the Information Society Showcase, held in conjunction with the G7 Ministerial Meeting in Brussels in February 1995. At the showcase, G7 ministers, specially invited delegates, and the attendant media were able to (*inter alia*):

- Share lessons with students around the globe in a "world classroom."
- Use a digital map that could be instantly updated over a mobile phone.
- Use a computer that could read electronic mail out loud.
- Dictate directly into a word processor.
- Work in two offices in different places at the same time.
- Control a TV with hand gestures.
- Walk through fire and handle hazardous objects in safety.
- Turn themselves into cartoon characters.[3]

Surely there was something there to catch the attention of any politician!

Another major initiative to raise awareness was the Global Knowledge Conference (GK97) held in Toronto in June 1997. Cosponsored by the World Bank and the Canadian government, the initiative for the conference was spearheaded by World Bank President James Wolfensohn, an almost evangelical convert to information and communication technologies (ICTs). The more than 2,000 attendees at GK97 could attend sessions on distance education, telemedicine, and other developmental applications of ICTs, as well as demonstrations and electronic testi-

---

1. G7: An association of the world's major industrialized economies: Canada, France, Germany, Italy, Japan, the United Kingdom, the United States.

2. Asia Pacific Economic Conference, an association of Asian and Pacific economies (the use of latter term, rather than *nations,* allows inclusion of entities such as Hong Kong and Taiwan).

3. Source: http://nii.nist.gov/g7/g7-gip.html

monials. For example, an Amazon tribal chief explained via teleconference how his village had struck a deal with British retailer Body Shops to sell essential oils from the rain forest.[4] Inuit children demonstrated a website where they publish videos of elders teaching traditional crafts; they also use the Internet for virtual frog dissection (real frogs are scarce in the central Arctic) and exchanging letters with children in other countries. The conference papers, available online, are also a valuable repository on ICTs and development.[5]

However, the biggest boost to awareness has been the Internet itself. Nontechnical decision makers such as Wolfensohn may not understand much about the Internet, but they know they cannot ignore it. They thus generally appear much more willing to embrace information technology initiatives than their predecessors.

JUMPING ON THE BANDWAGON: NATIONAL AND REGIONAL PLANS AND PROJECTS

Plans

The second step was preparing an NII or a regional II plan. Actually, many of the industrialized countries' plans were published before the G7 meeting, as their governments were spurred to action by the astonishing growth of the Internet and concern about the economic and technological lead the U.S. might gain through its commitment to an NII, as presented by the Clinton Administration in the 1993 NTIA report *National Information Infrastructure: Agenda for Action.* Notably, the Bangemann Report to the EU stated: "The first countries to enter the information era will be in a position to dictate the course of future developments to the late-comers."[6] (In fact, Singapore had adopted an II strategy in the 1980s through its commitment to becoming a regional information hub, a policy that evolved by 1992 into becoming an "intelligent island.")

Examples of II plans include:

- Singapore: *A Vision of an Intelligent Island (The IT 2000 Report)*, March 1992.
- United States: *The NII: Agenda for Action*, September 1993.
- South Korea: *Initiative for Building the Korea Information Infrastructure*, April 1994.
- European Union: *Europe and the Global Information Society* (Bangemann Report), May 1994.
- Japan: *Telecommunications Council Report*, May 1994.
- France: *Information Highway* (Thery Report), October 1994.

---

4. Wysocki (1997).
5. See http://www.globalknowledge.org/english/toronto97/index.html
6. High-Level Group on the Information Society (1994).

- Canada: *Canada's Information Highway*, November 1994.
- Finland: *Finland Towards the Information Society—A National Strategy*, December 1994.[7]

It is not necessary to do content analysis to find the similarity in these plans. It would be easy to substitute Canada or Australia or the United Kingdom for Finland in Finland's plan; the goals of South Korea, Taiwan, Malaysia, and the Philippines are also remarkably similar. Most recommend government support for II trials and projects. Some set specific targets such as Japan's fiber to all homes by 2010, Singapore's *IT 2000* and Malaysia's *Vision 2020*. Their strategies for extending or upgrading infrastructure, however, vary widely from Singapore's state investment in Singapore ONE, to government supported high-tech zones such as Malaysia's Multimedia Supercorridor and the Philippines' Subic Cybercity, to market-driven competition policies in Hong Kong, New Zealand, and the United States.

GII Projects

INDUSTRIALIZED COUNTRIES: THE G7

The next step was to develop projects and trials, as a means of demonstrating the potential benefits of ICTs. Table 10.1 lists the top 15 Internet user economies, all of which had more than 500 users per 10,000 population in 1996. It would appear that most GII applications in these countries, and others with high growth in Internet users, will be market driven. The role of pilot projects, then, would be to devise means to provide access to the disadvantaged, such as urban poor and isolated communities, and to apply ICTs to priority development goals such as improving educational standards, retraining the workforce, reducing medical costs, or reducing air pollution. It would also be interesting to determine what lessons might be learned from these countries about adoption of ICTs. What factors have driven such high penetration in Scandinavia and Australia? Why are other G7 nations (France, Germany, Italy, the United Kingdom) not in the top 15?

However, most of the 11 thematic areas selected by the G7 members seem to have little to do with understanding ICT adoption or applying information technology to high-priority problems. See Table 10.2.

These projects are at various stages of implementation. All appear to have encouraged collaboration among industry, academia, and the public sector; there is also extensive participation by G7 nations other than those responsible, and in many cases, by non-G7 nations. The sustainability implied by criteria of avoiding creation of new bureaucracies and obtaining funding from existing projects is uncertain. Some projects report that participants have had difficulty in finding funding; others have

---

7. Source: www.ncb.gov.sg/nii

initiated "free" trials that appear impossible to sustain. For example, the Progress Report on Broadband Networking notes: "Trans-oceanic interconnections are costly propositions for telecommunications carriers, especially for research, which by definition does not generate the revenue that commercial traffic does."[8]

TABLE 10.1
Top 15 Internet User Economies, 1996

Country	Internet users per 10,000 inhabitants
Finland	1,678.38
Bermuda	1,562.50
Iceland	1,482.98
Norway	1,138.17
Australia	1,092.18
Sweden	904.67
New Zealand	840.34
United States	787.32
Canada	667.48
Netherlands	580.01
Japan	556.61
Denmark	570.13
Luxembourg	557.17
Switzerland	521.13
Slovenia	502.26

Derived from International Telecommunications Union. *World Telecommunication Development Report 1998.* Geneva: ITU, 1998.

INTERNET INITIATIVES FOR DEVELOPING COUNTRIES

Despite massive investment in infrastructure in the developing world, there is still an enormous gap in access to telecommunications services between the industrialized countries and the poorest countries. However, at present, the gap in access to computers in developing countries is much greater than the gap in access to telephone lines or telephones. High-income countries had 22 times as many telephone lines per 100 population as low-income countries, but 96 times as many computers (see Table 10.3). In 1996, there were more than 100 times as many Internet users per 10,000 population in industrialized countries as in the poorest countries (see Table 10.4).

8. Source: http://nii.nist.gov/g7/g7-gip.html

TABLE 10.2
G7 GII Project Themes and Responsibilities

Theme	Responsible Nations
Global inventory of II projects	European Commission, Japan
Global interoperability for broadband networks	Canada, Germany, Japan, United Kingdom
Cross-cultural training and education	France, Germany
Electronic libraries	France, Japan
Electronic museums and galleries	Italy, France
Environment and natural resources management	United States
Global emergency management	Canada
Global health care applications	European Commission, France, Germany, Italy
Government online	United Kingdom, Canada
Global electronic marketplace for SMEs*	European Commission, Japan, United States
Maritime information systems	European Commission, Canada

Source: http://www.ispo.cec.be/g7/projidx.html.
* Small- and medium-sized enterprises.

TABLE 10.3
Teledensity and Computer Density, 1996

Country classification	Teledensity (lines/ 100 population)	PCs per 100 inhabitants
Low income	2.45	0.23
Lower middle income	9.71	1.34
Upper middle income	13.36	2.92
High income	54.06	22.28

Source: Derived from International Telecommunication Union. *World Telecommunication Development Report 1998*. Geneva: ITU, 1998.

Several international agencies are sponsoring projects to provide Internet access in developing countries, and others are emphasizing developmental applications of ICTs. Among them are the following:

- *Canada's International Development Research Centre (IDRC):* IDRC operates a global networking program consisting primarily of the Pan Asian Networking initiative that is now being expanded to cover other developing regions including Latin America, the Caribbean, North Africa, and the Middle East; and the Acacia initiative that is addressing the information needs of sub-Saharan Africa. IDRC assists developing country nongovernmental organizations and entrepreneurs in establishing Internet service providers, provides training, and emphasizes sustainability.[9]

- *USAID's Leland Initiative:* The Leland Initiative (named for deceased Congressman Mickey Leland), sponsored by the U.S. Agency for International Development (USAID), is a 5-year, $15 million U.S. government effort to extend full Internet connectivity to approximately 20 African countries to promote sustainable development. Leland has three major objectives:
- *Policies:* to create an enabling policy environment.
- *Pipes:* to foster a sustainable supply of Internet services.
- *People:* to promote Internet user applications for sustainable development.
- *UNDP's Sustainable Development Networking Program:* The Sustainable Development Networking Programme (SDNP), sponsored by the United Nations Development Program (UNDP), is a national information exchange operation run by independent entrepreneurs who receive equipment and seed funding from UNDP. A direct result of the 1992 United Nations Conference on the Environment and Development (the Rio Earth Summit), the SDNP links government organizations, the private sector, universities, nongovernmental organizations, and individuals through electronic and other networking vehicles to exchange critical information on sustainable development.[10]
- *The World Bank's InfoDev Project:* The World Bank points out that ICTs "open up extraordinary opportunities to accelerate social and economic development, and they create a pressing reform and investment agenda both to capitalize on the new opportunities and to avoid the deterioration of international competitiveness."[11] The InfoDev project aims to address this agenda by funding activities to assist developing countries and emerging economies to harness these technologies. Its strategies include leveraging funds and brokering partnerships to create a network for improved communication and information sharing.[12]
- *The Soros Foundation's Internet Program:* The Soros Foundation's Open Society Institute (OSI) is a private foundation that promotes the development of open societies around the world by operating and supporting a variety of initiatives and projects in education, independent media, legal reform, and human rights. The OSI's Internet Program supports projects to provide Internet access as part of its strategy for fostering open civil societies.[13]

Like the G7 projects, many of these developing country initiatives are in their early phases. It is interesting to note that the funding organizations tend to empha-

---

9. Source: www.idrc.ca
10. Lankester and Labelle (1997).
11. Source: www.worldbank.org/html/fpd/infodev/infodev.html
12. Ibid.
13. Source: www.soros.org

size project sustainability, and several include telecommunications policy reform as part of their agenda.

TABLE 10.4
Internet Access, 1996

Country classification	Internet hosts per 10,000 inhabitants	Internet users per 10,000 inhabitants
Low income	0.09	0.89
Lower middle income	1.93	19.00
Upper middle income	8.40	55.87
High income	28.14	91.89

Source: Derived from International Telecommunication Union. *World Telecommunication Development Report 1998*. Geneva: ITU, 1998.

CRITICAL ISSUES

The following sections examine critical issues that need to be addressed in implementing projects and trials of new communications technologies, using case study examples from current GII projects.

Technology and Free Tickets

The underlying assumption of the social and economic value of GII is that individuals and organizations will benefit from greater ability to access and share information. Yet, "information highways" initially provided enormous increases in bandwidth not in response to demand, but because of technological advances in network capacity and switching. The early phases of the II movement were primarily driven by the introduction of optical fiber; indeed, in the early 1990s, many pundits predicted the demise of satellites and other wireless systems for point-to-point communications. In this respect, the II movement resembled the experimental satellite era of the early 1970s when satellites were too often a solution in search of a problem, or, in some cases, a solution to a real problem that ignored the many other issues that had to be addressed to solve it. This approach goes beyond "Build it and they will come." It could be called "Build it and give away free tickets."

THE ALLURE OF BANDWIDTH

Many proponents of GII equate II with big pipes and fast switches, such as optical fiber with asynchronous transfer mode (ATM) switching. Following a technology-driven rationale, they then advocate support for trials and pilot projects that can take advantage of this capacity. Many of these projects involve research institutions with supercomputers, such as Pacific Bell's CalREN (California Research Network) and Canadian university projects sponsored by CANARIE. Health care trials typically focus on imaging such as CAT scans and MRI images sent between hospitals and specialists at major medical centers.

For example, Japan's Ministry of Posts and Telecommunications (MPT) has initiated telemedicine demonstration tests linking Kyushu University and the University of Occupational and Environmental Health in Japan utilizing fiber optic ATM communications. This trial is one of the first application experiments for the Global Interoperability for Broadband Networks project, one of the international joint projects selected at the 1995 G7 Ministerial Meeting on the Information Society. The trial is designed to diagnose intractable epileptics by exchanging, via a high-speed digital circuit, multichannel digital brain wave signals and synchronized video of patients experiencing an epileptic seizure.[14] If this is a valid telemedicine experiment, evaluation should include whether this problem is a high priority, how much bandwidth is actually required, and in this case, whether the researchers can actually transmit data while the patient is having a seizure, and whether real-time analysis is critical (otherwise slower data transmission or mailing of tapes would suffice).

MPT highlights the technology-driven rationale for such projects: "This experiment hopes to help Japan begin a full-scale broadband telemedicine test using fiber optics and showcase concrete examples of the advanced information society by developing the usage of the system." It further suggests that this rationale applies to G7 and APEC GII projects in telemedicine, noting that both Japanese universities plan a connection with the Cleveland Clinic Foundation in Ohio, described as "the world's first full-scale international broadband application test of its kind, which will contribute to the implementation of the Global Information Infrastructure (GII) and the information-networked society. . . . Furthermore, both universities hope to be active players in providing telemedicine applications throughout Asia. This project is expected to be a promising candidate for one of the Asia-Pacific Information Infrastructure (APII) test bed projects that were agreed upon in APEC meetings."[15]

In another telemedicine example, an EU telemedicine project called LOGINAT established an interactive video link between the Lille University Hospital Centre's (UHC) maternity department and its counterpart at the general hospital of Bethune, about 30 km away. The link allows specialists of the two hospitals to "meet" every week. The EU states that the Lille UHC "developed this project to allow quick decisions on whether children in urgent cases should be transferred to the UHC. The project allows a diagnosis to be made at a distance, including a visual

---

14. www.mpt.go.jp
15. www.mpt.go.jp

examination. . . . With UHC's help, Bethune can analyse difficult cases and decide on treatment through echographies and medical dossiers examined directly on a maxi-screen."[16] Note that the hospitals are less than 20 miles apart, and an interactive video link is specified. It would be important to know how frequently visual information is required to make such a determination, rather than transmission of data such as EKG, vital signs, blood analysis, weight, and so on, that would be part of the "medical dossiers," and require much less bandwidth. Also, why is fully interactive video required if the presentations are always from Bethune to Lille?

The EU summary on telemedicine projects goes on to state: "The quality of images limits the full development of telemedicine. The UHC is now trying to extend the videoconference system to two other hospitals of the Nord-Pas-de Calais region and to use it for permanent distance training of obstetricians." Again, how often is video required, and of what quality? Experience with distance education indicates that for adult education and training 384 kbps is often adequate, and even as little as 128 kbps may suffice. The latter is the equivalent of the two 64 kbps channels of an ISDN circuit; a few ISDN circuits, widely available in France thanks to the infrastructure investments of the 1980s, should be able to provide the necessary capacity for such applications.

In many cases, narrow bandwidth may suffice for medical applications. For example, transmission of patient data can be a cost-effective means of improving diagnosis and treatment. Health aides in Alaska have used satellite audio channels for 20 years to consult with regional physicians. Such channels can be used to transmit chest sounds, EKG tracings, and other patient data. U.S. trials have shown that patients may also be discharged sooner if they can be monitored at home with data transmitted over their telephone lines. The electronic availability of patient records, which requires very little bandwidth, could have a significant impact on efficiency of patient treatment and reduced guessing and errors in treating patients with unknown medical histories. The EU supports several telematics initiatives in electronic patient data, including the development of medical smart cards.[17]

In developing countries, a recent project involving specialists at Yale University and physicians in rural Ecuador used laptop computers and dial-up phone lines to link a mobile surgery program in the Andes with an isolated hospital 10 hours away and with a specialist at Yale. The Ecuadorian doctors in Cuenca examined their patients in Sucua and discussed their care with primary physicians using phone lines; they also performed a laparoscopic procedure with real-time consultation and monitoring from the Yale campus.[18] In Africa, the Healthnet project has used a simple store-and-forward satellite data circuit to collect epidemiological data and to enable African physicians to access the National Library of Medicine

---

16. Source: www2.echo.lu/eudocs/en/eudocshome.html
17. Source: www2.echo.lu/eudocs/en/eudocshome.html
18. Source: http://yalesurgery.med.yale.edu/

and communicate with each other via e-mail. This latter function of information sharing, now expanding in developing countries via Internet access, may be the most valuable of all medical applications, as there have been examples of physicians in China and Africa successfully treating undiagnosed patients after seeking advice over the Internet.[19]

Where Are the Users?

One reason that many initiatives seem overwhelmingly technology-driven is that they are being implemented primarily by technocrats. For example, in Asia, the Asia-Pacific Telecommunity (APT) has identified APII projects as a priority. At its High Level Development Meeting on APII, not only were there no representatives of users present, but strategies for involving users or identifying priorities for applications were not discussed. Instead, each country, represented by its telecommunications administration, reported on its progress in developing APII initiatives. The regional projects were dominated by Japan and Australia, the major funding partners and sources of expertise. No effort was made to facilitate exchange of experiences among countries with similar problems or shared development agendas, such as the Pacific Island nations or members of the Association of Southeast Asian Nations (ASEAN). Instead, countries simply reported over 2 days in alphabetical order, with China followed by the Cook Islands, and Thailand by Tonga.[20] A much more user-inclusive approach is being used in Hong Kong, where the Office of the Telecommunications Authority has established an Information Infrastructure Advisory Committee, with a Task Force on Applications and working groups on education, business, government, the community, the environment, and personal services.[21] Workshops have enabled the public to gain hands-on familiarity with ICTs.

Even when users are involved, the availability of funds for NII and GII projects can skew priorities. As was true with the experimental satellite projects of the 1970s and 1980s, projects may be proposed because funds are available, rather than because they may solve a high-priority problem such as reducing air pollution in Asian cities or upgrading the skills of rural teachers and health care workers. An offer of free hammers tends to generate proposals for projects that look like nails, even if nails are not really a priority.

Incentive-Based Strategies

Without incentives and links to high-priority problems, II projects and trials are likely to terminate when funding ends or the free ride on the network is over, even

---

19. Source: www.healthnet.org
20. The author attended the Bangkok APT meeting on APII in June 1997 as an observer.
21. Source: http://www.ofta.gov.hk/index_eng1.html

if they have demonstrated successful applications of the technology. Such was the case of many of the early North American satellite experiments. In 1977, Dr. Maxine Rockoff, who had supported several telemedicine projects on the joint NASA/ Canadian Communications Technology Satellite (CTS) through the National Center for Health Services Research, pointed out that even successful telemedicine trials were not being continued, apparently because the U.S. health care system at that time had no incentive to reduce the number of referrals from general practitioners to specialists or to fill fewer hospital beds.[22] Her speech was widely criticized by defensive NASA officials.

Rather than starting with technology, a more developmentally sound approach would be to examine the nation's or region's social and economic priorities, typically spelled out in government policy documents, and then to postulate where applications of information technology could significantly contribute to achieving the goal or addressing the problem. Incentives from the sectors themselves may spur them to try II-based solutions. For example, Texas reformed its high school curriculum in the early 1980s, requiring students to take a foreign language and more advanced math and science courses. Rural schools had a problem: If they could not offer the newly required courses, their students would be bussed to larger schools, and the local schools could be closed. Many rural school districts did not have the funds to attract specialized teachers; even if they could find money, they found it difficult to attract such teachers to their small communities. To address this problem, an educational satellite network called TI-IN was established to deliver the courses. Today, TI-IN serves rural schools in 46 states.[23]

Another incentive-driven organization is the National Technological University (NTU), which delivers graduate technical courses to high-tech companies via satellite. NTU is actually a consortium of universities that offer local courses over telecommunications networks such as Instructional Technology Fixed Service (ITFS) and cable; the demand to offer these courses nationally came from employers who wanted their employees to stay up to date in their fields without having to commute to campus or take long leaves of absence to pursue advanced degrees.

In California, new environmental controls on air pollution may create incentives for employers to encourage telecommuting because they will be forced to get more of their employees off the roads. Part of the solution could be to let more employees work either from home or from telecenters close to their residences. Many of Southeast Asia's sprawling cities such as Bangkok, Jakarta, Kuala Lumpur, and Manila are notoriously polluted, primarily from automobile exhaust. Better public transit systems are desperately needed; the establishment of telecenters near residential areas could also help by enabling information workers ranging from data entry clerks and typists to engineers to work close to home. However, successful

---

22. Rockoff, quoted in Hudson (1990, p. 58).
23. Parker and Hudson (1995).

implementation of such a strategy will require not only infrastructure but also incentives to reduce pollution that do not exist at present.

Sustainability

A flaw in many of the experimental projects of the 1970s and 1980s was that there was no planning or transition strategy to continue the application after the experimental period. Most of the NASA Applied Technology Satellite (ATS) and joint U.S.–Canadian CTS projects died when the experimental period was over. Some trials that were positively evaluated (e.g., those of the Veterans Administration) ended because there was not a strong enough commitment to them to allocate the necessary funding from organizational budgets. Others, such as the participants in PEACESAT, which originated on ATS-1, struggled to find free satellite time, eventually on other satellites. (Two exceptions, where commitment was coupled with successful entrepreneurial and policy strategies, were the ATS-1 health network in Alaska that led to satellite-based telephone service for Alaskan villages, and the Appalachian ATS-6 project that eventually became the Learning Channel.[24])

These trials were on so-called experimental satellites that were not authorized to provide operational services, so that experimenters had to find other satellite capacity as well as funds to pay for it if they wanted to continue. A more comparable example to today's NII and GII trials was Intelsat's Project SHARE, which provided free satellite time for distance education and telemedicine experiments on operational Intelsat satellites. Significantly, the applications died when the free time expired; Project SHARE did not result in operational follow-on services or new revenue for Intelsat. A more market-driven successor called Project ACCESS had very limited success in generating business for Intelsat.

The II experiments and trials of the late 1990s typically use operational networks, typically built by telecommunications carriers, although some may use dedicated government networks. Most utilize fiber optic networks; even if the networks are not profitable, they are not going to be ripped out of the ground at the end of the experiments. Thus, experimenters do not have to look around for other capacity (although other technologies such as cable and satellites may turn out to be attractive alternatives), but they do have to find the funding to pay for use of the network and continued operation and maintenance of the equipment they need for the application.

Transitional strategies for the GII projects are unclear. Will Japan's MPT continue to pay for links from Japanese universities to the United States and Southeast Asia? Will EU member governments provide funds in their budgets to pay the now commercialized operators in their countries for connections among hospitals,

---

24. See Hudson (1990).

schools, training centers, and so on? Will discounted rates for educational or other social service users be required by regulatory bodies?

The guidelines for U.S. subsidy programs for Internet connection for schools mandated by the 1996 Telecommunications Act may be provide a useful model for other countries. Before applying for a subsidy, the school must file an approved technology plan that states how it intends to use the Internet connection and how it plans to train teachers, purchase and maintain equipment, and pay its share of the Internet connection.[25] The plan is required to ensure that schools do consider and plan for the real costs and responsibilities of network access in an attempt to avoid past education technology project failures in which equipment collected dust because users were not trained or applications were not considered relevant, or projects were abandoned when funds ran out.

The Need for Evaluation

Notably missing from the list of criteria for G7 projects is any requirement for evaluation. Surely, the most important benefit of these projects would be an evaluation of the extent to which they achieved their goals and met the preceding criteria, particularly "adding value for developing the Information Society" and sustainability. Operationalizing these criteria and their internal project goals and collecting and disseminating evaluation findings, as opposed to descriptive reports by project staff, could benefit not only the participants and their funding sources, but others interested in these applications. In this respect, the G7 projects, and many NII projects, differ from the North American satellite experiments of the 1970s (on ATS-1, ATS-3, ATS-6, and CTS) that required rigorous external evaluations.

Evaluation is not, or should not be, simply an interesting research opportunity. Lessons learned may prove useful for other similar applications and may also be important in determining the future of the project, and possibly the whole initiative. An example is the current proliferation of telecenter projects being funded by agencies including the World Bank, ITU, UNESCO, IDRC, and other organizations. It will be critical to gather data to determine what factors determine how such facilities are used and by whom, what training is needed to operate the centers, what strategies may be effective in ensuring sustainability past the pilot stage, and, most significantly, how and to what extent the ability to access and share information through these centers can contribute to social and economic development.

Social Versus Economic Goals: "When the Door Opens ..."

Citizen and consumer organizations, religious groups, political parties, and national governments in many countries have voiced concerns about the potential

---

25. See www.slcfund.org and www.eratehotline.org

perils of unfettered access to information available on the Internet. Proposed options to limit access range from "Net nanny" software to site certification by various bodies to requirements by governments for Internet service providers to block access to offensive sites. An analysis of these options and their implications is beyond the scope of this chapter.[26] However, it is important to point out that in many countries there is a tension between economic and social goals for increasing access to the GII.

This tension is most evident in Asian countries that identify investment in their NII as a means to make their own economies attractive for foreign business and to develop information technologies for export; examples of such countries include China, Malaysia, South Korea, and Singapore. Chinese premier Deng Xiaoping expressed this ambivalence about opening China's doors to the world: "When the door opens, some flies are bound to come in."[27] China is greatly expanding access to the Internet through investments in telecommunications and authorization of private service providers, but continues to block access to Internet sites and ban satellite antennas.

However, it is Singapore where the policies, at least from a Western perspective, are the most contradictory. Singapore has deliberately attracted foreign satellite uplink operators to locate in its high-tech industrial parks, but bans satellite terminals that could receive these channels or others available on the many satellites with footprints covering the country.[28] Singapore's most recent commitment to becoming an "intelligent island," inaugurated at the ITU's Asia Telecom 97 (held in Singapore), is Singapore ONE, a national initiative to deliver a new level of interactive, multimedia applications and services to homes, businesses, and schools throughout Singapore. Singapore ONE is comprised of two components: a broadband switched network and advanced applications and services designed to take advantage of the infrastructure's high-speed and high-capacity capabilities.[29]

By the end of its first year, Singapore ONE had begun to establish "Smart Clubs" for users. The Singaporean vision of access with control is evident from the website description of the model Smart Club, which is to have the following features:

- A Webcam in which users can remotely observe the activities at the Singapore ONE Club. "This is useful for parents at home whose child is using the computer facilities at the Club, or for teachers to check on the students' progress on the computer."
- Remote operation where the Club's doors can be opened by the authorities remotely.

---

26. See Lessig and Resnick (chap. 9, this volume).
27. Schwankert (1995).
28. See Hudson (1997b).
29. See www.ncb.gov.sg

- Remote monitoring in cases where the authorities may need to terminate access to undesirable sites by the users.
- Usage tracking so that the Club can record the frequency and usage habits of users. "This will help the Club to understand the usage preferences and thus enable it to plan its programmes and activities to better meet the needs of its members."
- A quick recovery system in which the system can be brought back quickly into operation and reinstalled if necessary if an outsider intrudes or hacks into it.
- Smart cards that allow users to access the Club's premises. "In the near future, the smart card will also become an access card to enable members to log on to the computers as well as act as a membership card."[30]

This description provides a revealing glimpse of the Singaporean vision of telecenters, with many forms of electronic monitoring in Smart Club packaging. Other organizations may not implement such surveillance in their II projects, but monitoring users through smart cards and servers as well as controlling access to information will certainly be feasible and probably will not be perceived by most users.

CONCLUSION: BEYOND TECHNOLOGY

Despite their ostensible socioeconomic objectives, many GII initiatives are technology driven, with the major instigators being the technical ministries, telecommunications operators, and equipment suppliers. Applications projects too often appear as afterthoughts, planned to take advantage of the facilities rather than to meet the needs of educators, health care providers, or nongovernmental organizations. These information highway initiatives assume that converging technologies will result in information services with both social and economic benefits. However, this assumption needs to be carefully examined. Each new communication technology has been heralded as offering numerous benefits. Satellites and cable television were to provide courses taught by the best instructors to students in schools, homes, and workplaces. Videoconferencing was to largely eliminate business travel. Telemedicine was to replace referral of patients to specialists. Computers were to replace traditional teaching with more personalized and interactive instruction.

To some extent all of the prophesies have been fulfilled, yet the potential of the technologies is far from fully realized. In many cases, it took institutional change and incentives to innovate for these technologies to have much effect. Now telecenters are being promoted as a means of providing access to communications technologies and services that could contribute to community development. Yet,

---

30. See www.ncb.gov.sg

as has been learned from experience with the previous innovations, investment in technology alone will not likely result in major social benefits.

To make a contribution to our understanding of how new technologies can contribute to social, economic, and cultural development goals, GII initiatives must at minimum:

- Be designed to contribute to priority goals, rather than being driven exclusively by technology.
- Take into consideration the local environment in terms of availability of equipment, skills, and other resources.
- Involve the users.
- Include external evaluation.
- Plan for sustainability past the project period.

Although policymakers may be aware that public sector stimulus is needed to foster applications of new technologies, support for trials and pilot projects may not ensure long-term implementation. If the services are perceived as frills diverting energy and resources from higher priorities, or if there is no plan to sustain operations past the pilot phase, the free ticket will have bought only an expensive excursion.

REFERENCES

High-Level Group on the Information Society. (1994). *Europe and the global information society: Recommendations to the European Council* (The Bangemann Report). Brussels, Belgium: European Commission.

Hudson, H. E. (1990). *Communication satellites: Their development and impact.* New York: The Free Press.

Hudson, H. E. (1997a). *Global connections: International telecommunications infrastructure and policy.* New York: Wiley.

Hudson, H. E. (1997b). Restructuring the telecommunications sector in developing regions: Lessons from Southeast Asia. In G. L. Rosston & D. Waterman (Eds.), *Interconnection and the Internet: Selected papers from the 1996 Telecommunications Policy Research Conference* (pp. XXX–XXX). Mahwah, NJ: Lawrence Erlbaum Associates.

International Telecommunication Union. (1998). *World telecommunication development report 1998.* Geneva, Switzerland: Author.

Kahin, B., & Wilson, E. J., III (Eds.). (1997). *National information infrastructure initiatives.* Cambridge, MA: MIT Press.

Langdale, J. V. (1998). The national information infrastructure in the Asia-Pacific Region. In *Proceedings of the 1998 Pacific Telecommunications Conference.*

Lankester, C., & Labelle, R. (1997, June). *The Sustainable Development Networking Programme (SDNP): 1992–1997.* Paper presented at the Global Knowledge Conference, Toronto.

Parker, E. B., & Hudson, H. E. (1995). *Electronic byways: State policies for rural development through telecommunications* (2nd ed.). Washington, DC: Aspen Institute.

Schwankert, S. (1995). Dragons at the gates. *Internet World,* November, p.112.

U.S. Department of Commerce. (1993). *National information infrastructure: Agenda for action.* Washington, DC: Author.

U.S. Department of Commerce. (1995). *Global information infrastructure: Agenda for cooperation.* Washington, DC: Author.

Wysocki, B., Jr. (1997, July 7). Development strategy: Close information gap. *The Wall Street Journal,* Dow Jones Online News.

WEBSITES

http://nii.nist.gov/g7/g7-gip.html
http://yalesurgery.med.yale.edu/
www.eratehotline.org
www.globalknowledge.org/english/toronto97/index.html
www.healthnet.org
www.idrc.ca
www.info.usaid.gov/regions/afr/leland
www.ispo.cec.be/g7/projidx.html
ww.mpt.go.jp
www.ncb.gov.sg/nii
www.ofta.gov.hk/index_eng1.html
www.slcfund.org
www.soros.org
www.worldbank.org/html/fpd/infodev/infodev.html
www2.echo.lu/eudocs/en/eudocshome.html

III

UNIVERSAL SERVICE

The Persistent Gap in Telecommunications: Toward Hypothesis and Answers

Jorge Reina Schement
Scott C. Forbes
Pennsylvania State University

In this chapter we identify demographic factors most closely identified with the gap in household information technology (IT) penetration—particularly the telephone—among Whites, AfricanAmericans, and Hispanics. Not surprisingly, income is the dominant factor affecting ownership and use of IT, but gender, age, marital status, and regional housing characteristics are also contributing elements to the IT race and ethnicity gaps. By analyzing underlying patterns of U.S. telephone penetration inequality at the national, state, and local levels, we describe the characteristics of these persistent gaps and propose recommendations for affecting and redefining U.S. universal service policies. Some of these suggestions include using data that specifically describe penetration gaps below the national level and recognizing that most IT gaps are amalgamations of smaller socioeconomic trends that can be discerned and reduced only with careful historical analysis of both technology choices and living patterns. Indeed, penetration gaps in the United States are affected by a complex array of factors more particular to localities than to the country as a whole. As such, they must be investigated with rigor and caution.

INTRODUCTION

I swear to the Lord,
I still can't see,
Why Democracy means,
Everybody but me.
        —Langston Hughes[1]

Does this gap in access to technology matter? You bet it does. How can you look for a job without a phone? How can an employer call you for an interview? How can you demonstrate that you have the skills to compete if you don't know which side of a diskette goes in first?

—William E. Kennard[2]

Langston Hughes's plaintive lament strikes at the heart of the intention of democracy. Democracy's ideal—sovereignty and inclusion—thrives or withers on democracy's reality. To argue that a democracy can exist when its reality excludes individuals who are nevertheless bound by its laws is at best to claim a democracy in the making and at worst to live an enduring hypocrisy. For the most part, Americans have claimed a democracy in the making and have struggled for the last 200 years to establish genuine participation for all citizens. Indeed, we tacitly acknowledge that for democracy to live up to its ideal, it must include all of its members from the core to the periphery. The key to the struggle and the promise has been and continues to be participation.

In the information age, universal access to communications technology is the primary policy tool for enabling citizens to participate in the economic, political, and social activities fundamental to a well-functioning and stable society. Therefore, Federal Communications Commission Chairman William E. Kennard's questions engage far more than a concern for opportunities available to Blacks or Hispanics. Kennard was addressing the promise of democracy, just as Langston Hughes addressed the shame of its blindness.

In this chapter, we concentrate on observed household penetration gaps in IT among White, Black, and Hispanic Americans.[3] We ask what the characteristics are of the persistent gaps in ITs and telecommunications networks among the na-

---

1. Langston Hughes (1943). The black man speaks. In *Jim Crow's last stand*.

2. William Kennard, Chairman of the Federal Communications Commission, Speech to the National Urban League, Philadelphia, August 3, 1998.

3. We use the terms *White, Black,* and *Hispanic* mainly because the U.S. Census Bureau uses these classifications for technology penetration data. White and Black are considered racial categories. Hispanic is considered an ethnic category; it can incorporate persons belonging to multiple racial categories. For example, a *Hispanic* could be a White Hispanic, a Black Hispanic, or an Asian Hispanic. More specifically, *Black* characterizes persons who identify themselves as Black, Negro, African American, Jamaican, Nigerian, West Indian, or Haitian. Hispanic categorizes persons of any race who identify themselves as Spanish, Spanish American, Latino, Mexican, Puerto Rican, Cuban, and of Central or South American descent. *White* characterizes people of European descent, such as those from Germany, France, Scotland, Poland, and so on.

tion's majority Whites and minority Blacks and Hispanics. Our purpose is thus to address this question and propose directions for arriving at an answer.

SOME CAVEATS

This chapter is an exploratory analysis of household penetration of ITs and services, specifically radio, television, telephone, cable, personal computers, and Internet access. We use descriptive statistics and focus on regional and ethnic comparisons in lieu of regression analyses in our discussion of the gaps to capture the widest possible audience and avoid convergence and possible confusion with the cost proxy model literature; a more exhaustive analysis of the data is in progress.

We do not discuss the admittedly important issues surrounding the delivery of universal service to libraries and schools or the gaps in equity of education that may result from differential levels of institutional access. Such policy questions more properly belong within a discussion of the necessary development of an institutional infrastructure to match the growth of the technological information infrastructure; they therefore stand beyond the scope of this chapter.

TELECOMMUNICATION GAPS IN 20TH CENTURY AMERICA

The literature on telecommunications gaps is small. Of the hundreds of studies concerned with universal service, only a few address telephone penetration and its social causes. Until the late 1980s the weight of opinion on households without telephones seems to have been that existing subsidy programs adequately included all those that could reasonably be connected.[4] Thus, for the century of telephone service, little or no thought was given to those left off of the national network. Only in the period immediately after the breakup of AT&T did some voices focus on the social dynamics of those without telephone service and point to poverty-related factors as causes of phonelessness.[5] These studies constitute the empirical source of the call to rethink universal service accomplishments in light of the emergence of a new information infrastructure. Finally, recent research indicates that those at the margins of society are particularly vulnerable to isolation and its socioeconomic consequences as a result of lack of telephone access.[6] However, al-

---

4. See Booker, E. (1986). Lifeline and the low income customer: Who is ultimately responsible? *Telephony, 210*(20), 116–132; Dordick, H. S. (1990). The origins of universal service. *Telecommunications Policy, 14*(3), 223–238; Dordick, H. S., & Fife, M. D. (1991). Universal service in post-divestiture USA. *Telecommunications Policy, 15*(2), 119–128; Gilbert, P. (1987). *Universal service on hold: A national survey of telephone service among low income households,* No. U.S. Public Interest Research Group; Hills, J. (1989). Universal service: Liberalization and privatization of telecommunications. *Telecommunications Policy, 13*(2), 129–144.

though some of these studies—two most recently from the government—point to and describe the gap in telephone penetration between the majority and minorities, there is almost nothing that points to causes of the gap.[7]

Under these circumstances, an analysis of historical gaps can be instructive. The spread of some earlier ITs indicates that gaps can be temporary.

In 1925, 10% of all households owned radios (Fig. 11.1). By 1930, ownership stood at 46%. Ten years later, in the throes of the Depression, Americans still managed to increase ownership of radios to 82% of all households. They bought radios at an astonishing rate, especially when one considers that the Depression forced personal expenditures on information goods and services to drop from 4.4% of all personal expenditures in 1930 to 3.5% in 1935, not recovering the 1930 level until 1945.[8] Despite these obstacles, radio achieved virtual saturation by 1950. Radio's astonishing growth masks the existence of a gap that might have existed between

---

5. See Perl, L. J. (1983). *Residential demand for telephone service 1983* (No. 1). National Economic Research Associates, Inc. for the Central Services Organization, Inc. of the Bell Operating Companies; Hausman, J., Tardiff, T., & Belinfante, A. (1993). The effects of the breakup of AT&T on telephone penetration in the United States. *The American Economic Review, 83*(2), 178–184; Schement, J. R. (1994). *Beyond universal service: Characteristics of Americans without telephones, 1980–1993* (Communications Policy Working Paper No. 1). Benton Foundation; Schement, J. R. (1998). Thorough Americans: Minorities and the new media, *Investing in Diversity: Advancing Opportunities for Minorities and the Media.* Washington, DC: Aspen Institute; Schement, J. R., Belinfante, A., & Povich, L. (1997). Trends in telephone penetration in the United States 1984–1994, In E. M. Noam & A. J. Wolfson (Eds.), *Globalism and localism in telecommunications* (pp. 167–201). Amsterdam, The Netherlands: Elsevier; Mueller, M., & Schement, J. R. (1996). Universal service from the bottom up: A study of telephone penetration in Camden, New Jersey, *The Information Society, 12,* 273–292; Schement, J. R. (1995). Beyond universal service: Characteristics of Americans without telephones, 1980–1993, *Telecommunications Policy, 19*(6), 477–485; Williams, F., & Hadden, S. (1991). *On the prospects for redefining universal service: From connectivity to content.* Austin: Policy Research Project, The University of Texas; Williams, F., & Hadden, S. (1992). On the prospects for redefining universal service: From connectivity to content. *Information and Behavior, 4,* 49–63.

6. See Schement, J. R. (1998). Thorough Americans: Minorities and the new media. *Investing in diversity: Advancing opportunities for minorities and the media.* Washington, DC: Aspen Institute; Schement, J. R., Belinfante, A., & Povich, L. (1997). Trends in telephone penetration in the United States 1984–1994. In E. M. Noam & A. J. Wolfson (Eds.), *Globalism and localism in telecommunications* (pp. 167–201). Amsterdam, The Netherlands: Elsevier; Schement, J. R. (1995). Beyond universal service: Characteristics of Americans without telephones, 1980–1993. *Telecommunications Policy, 19*(6), 477–485.

7. *Falling through the Net: A survey of the "have nots" in rural and urban America.* (1995). Washington, DC: National Telecommunications and Information Administration, U.S. Bureau Of The Census, U.S. Department Of Commerce. *Falling through the Net II: New data on the digital divide.* Washington, DC: National Telecommunications and Information Administration, U.S. Bureau Of The Census, U.S. Department Of Commerce.

the majority and minorities. If there ever was a gap, and given what we know of subsequent gaps there is every reason to believe that a radio gap existed, that gap closed by 1950 so that nearly every household had a radio as they still do today.

FIGURE 11.1
Household Penetration of Selected Media, 1920–1998.

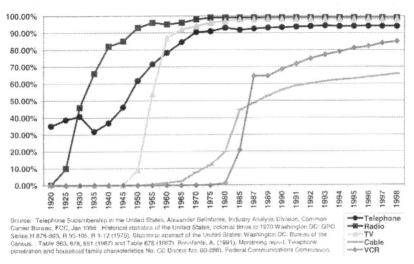

Source: Telephone Subscribership in the United States, Alexander Belinfante, Industry Analysis Division, Common Carrier Bureau, FCC, Jan 1998. Historical statistics of the United States, colonial times to 1970 Washington DC: GPO Series H 878-893, R 90-106, R 1-12 (1975). Statistical abstract of the United States: Washington DC: Bureau of the Census. . Table 363, 878, 951 (1987) and Table 676 (1997) Reinfante, A. (1991), Monitoring report. Telephone penetration and household family characteristics No. CC Docket No. 80-286), Federal Communications Commission.

Telephone
Radio
TV
Cable
VCR

In 1950, less than 1 household in 10 owned a television. However, 15 years later, less than 1 household in 10 remained without a TV. Television's complete adoption took less time than radio. Yet in this case, we have evidence of a gap during the period of television's diffusion. During the first 7 years of the diffusion of television, lower income groups lagged behind wealthier groups. The wealthiest quartile led the way so that by 1956 this group had reached 90% penetration. Nevertheless, by 1970, television had reached a saturation level close to that of radio. Again, as with radio, a gap closed.

The third IT of the era, the telephone, presents a stark contrast. From 1878, with the establishment of the first practical exchange, 80 years passed before three out of four households boasted a telephone.[9] Although the adoption of radio sets proved immune to the Depression, telephone penetration dipped in correlation

---

8. Series E 135–166, G 416–469 (1975). *Historical statistics of the United States, colonial times to 1970.* Washington, DC: U.S. Government Printing Office. Table 708, 738 (1981). *Statistical abstract of the United States: 1981.* Washington, DC: U.S. Bureau of the Census. Table 676 (1987). *Statistical abstract of the United States: 1988.* Washington, DC: Bureau of the Census.

with personal expenditures. Telephones reached saturation by 1970, with 93% of households slowly advancing to 94% in the years since.[10]

The key to the contrast lies in the nature of the exchange. Radio and television constitute goods, whereas the telephone constitutes a service. Telephone services require a decision to pay a monthly fee and the building of an infrastructure for the connection to function. For households on the margin, the payment structure of telephone service means a hard choice every month. As recent research suggests, that choice may result in the rejection of telephone service in favor of other purchases deemed more essential.[11] By comparison, the one-time cost characteristic of radio and TV (allowing them to circulate second- and thirdhand) facilitated rapid diffusion, obstructed the backslide experienced by the telephone during the Depression, and closed whatever gaps might have existed during radio and television's early diffusion. Telephone's gap still exists 120 years later. The lesson is that information goods have tended to diffuse more rapidly than information services. Thus, those gaps that may have existed in the initial distribution of radio and television closed, but the gap for telephone service did not.

DEMOGRAPHICS OF THE TELEPHONE GAP IN DETAIL

A review of the factors contributing to nonsubscribership indicates that households without phones comprise multiple but overlapping groups.

- Phone penetration in the suburbs is roughly 4.6% higher than in the central cities, and 4.1% higher than in households outside of metropolitan statistical areas (MSAs). Moreover, since 1984, the biggest increases in penetration have been in nonmetropolitan areas (not in MSAs), 89.2% to 92.5%.[12] Blacks and Hispanics tend to live in urban areas, so would seem to be affected negatively by the "urban handicap."
- Income makes a difference. Households receiving energy assistance,

---

9. Brooks, J. (1975). *Telephone: The first hundred years.* New York: Harper & Row, p. 65.

10. Belinfante, A. (1993, July). *Telephone subscribership in the United States* (CC Docket No. 87–339). Washington, DC: Federal Communications Commission.

11. Horrigan, J., & Rhodes, L. (1995). *The evolution of telephone service in Texas* (Working Paper). LBJ School of Public Affairs; Mueller, M. L., & Schement, J. R. (1996). Universal service from the bottom up: A study of telephone penetration in Camden, New Jersey. *The Information Society, 12,* 273–292; Schement, J. R., Belinfante, A., & Povich, L. (1997). Trends in telephone penetration in the United States 1984–1994. In E. M. Noam & A. J. Wolfson (Eds.), *Globalism and localism in telecommunications* (pp. 167–201). Amsterdam, The Netherlands: Elsevier.

12. Between 1985 and 1986, the Census changed definitions for "MSA status not identifiable." Many of those households were moved into "City status in MSA not identifiable," but this does not change the overall tendency.

food stamps, school lunch programs, welfare, and public assistance tend to have lower telephone penetration rates, as much as 20 points below the national average.[13] As minorities are more likely to have incomes below the national median, they will also experience the effects of income on telephone service with greater intensity than the population as a whole.

- Housing characteristics also influence telephone penetration. For families in multiple-unit housing and in rental housing, telephone penetration is lower (84.4% for occupants of rental housing). Here, too, minorities are more likely to experience these conditions, especially given their propensity for urban dwelling.

- Women experience phonelessness in inordinate numbers. The Census category "single civilian female with children" has the second-lowest telephone penetration (82.6%), exceeded only by the homeless. In addition, this category has shown little improvement (80.1% in 1984 to 82.6% in 1993). Here it would seem that Black women may be overrepresented, thus contributing to the gap.

- Youth suffers as well. Younger households—Whites, Blacks, and Hispanics—suffer lower telephone penetration levels than do households headed by older people. In November 1983, telephone penetration in Black households headed by 15- to 24-year-olds stood at 49.9% (these households being headed by women, for the most part). By 1988, penetration had risen to 65.6%, an increase of 31%. Between 1988 and 1992, the penetration curve flattened, and then turned up slightly in 1994 (74.6%). In White households headed by 15- to 24-year-olds, the 1983 level was 76.6%; in 1994, penetration stood at 86.3%. Telephone penetration among young Hispanic households began at 71.9% in 1988 and fluctuated to 73.9% in 1991, after which a significant increase took place, evening out at around 77% in 1994. There is clearly a gap visible here, but it is confounded by age.

- Finally, unemployment directly affects telephone penetration. All groups experienced lower penetration among the unemployed: Whites, 89.9%; Blacks, 77.9%; Hispanics, 85.6%.[14] Yet here too we see a 10-point spread between Whites and Blacks. It is reasonable to surmise that Blacks are hit harder by unemployment, but it is equally reasonable to surmise that unemployment does not stand alone as a factor.

13. Schement, J. R., Belinfante, A., & Povich, L. (1997). Trends in telephone penetration in the United States 1984–1994. In E. M. Noam & A. J. Wolfson (Eds.), *Globalism and localism in telecommunications* (pp. 167–201). Amsterdam, The Netherlands: Elsevier.

14. Schement, J. R., Belinfante, A., & Povich, L. (1997). Trends in telephone penetration in the United States 1984–1994. In E. M. Noam & A. J. Wolfson (Eds.), *Globalism and localism in telecommunications* (pp. 167–201). Amsterdam, The Netherlands: Elsevier.

In general, we can say that there is a positive correlation between income and telephone penetration, but income does not operate in a vacuum: Other factors compound or lessen the income effect. Households receiving any kind of government assistance fall below national telephone penetration levels. Likewise, women heads of households with children fare poorly. The unemployed suffer inordinate loss of telephone service. Through it all, minorities, especially Blacks, fall to the bottom of nearly all categories.

In short, these six contributing factors illustrate the interwoven complexity of access to a technology most people think of as simple: The phoneless do not constitute a homogeneous group. Any individual without a phone is likely to fall into several of these groups. To that extent, understanding the causes of low telephone penetration for any one group requires parsing out multiple contributing factors, something that is virtually impossible with our current statistics.

ETHNICITY AND THE TELEPHONE GAP

Minority household penetration rates consistently rank 8 to 10 percentage points lower than their White counterparts. In 1984, shortly after AT&T's modified final judgment, the gap between White and minority households was approximately 13%. Fifteen years later, White households have a national average telephone penetration rate of 95%, approximately 8% higher than the Black and Hispanic rates of 86.9% and 86.7%, respectively (Fig. 11.2).[15]

Differentials also exist among minority groups; a telephone penetration gap exists between Blacks and Hispanics, even adjusted for tenure and income. For example, in 1990 Black renter-occupied housing units in Pennsylvania with median incomes below the poverty line experienced a telephone penetration rate of 84%, whereas the penetration rate for Hispanics in similar circumstances was 72%— startlingly low numbers considering that in 1990 Pennsylvania had among the highest telephone penetration rate in the nation at 97.1%. In nearby Delaware, a state that also had a high telephone penetration level (95.7%), the magnitude of the 1990 gap was reversed: Hispanics had a 62% penetration rate and Blacks contended with a 55% penetration rate.[16]

Native American[17] households, particularly those on reservations and trust lands, experience the widest gap between their levels of telephone penetration and

15. Belinfante, A. (1998). *Telephone subscribership in the United States.* Washington, DC: Federal Communications Commission.

16. 1990 Census Summary Tape File 3.

17. According to the Census Bureau, persons who broadly identify themselves as Native American, Eskimo, or Aleut. More specifically, the term includes any indigenous people of the United States. Native American tribes are aggregated into the term *Native American.* Native American is considered a racial category.

both the White and overall national penetration averages. The two states with the largest numbers of Native Americans—Oklahoma and Arizona—rank in the bottom quintile of overall state telephone penetration. The 314 U.S. reservations and trust lands have an average telephone penetration rate of 46.6%, less than half the national average. Native Americans in the Southwest suffer particular hardship. For instance, the Navajo Reservation and accompanying Trust Lands in Arizona, New Mexico, and Utah suffer an astounding telephone penetration rate of 18.4%. That is, four of five American Indians in these territories do not have a phone. On the 48 reservations with 500 or more American Indian households, only 6 had telephone penetration levels above 75% and none exceeded 85%.[18]

FIGURE 11.2
Telephone Penetration by Ethnicity, 1983–1997.

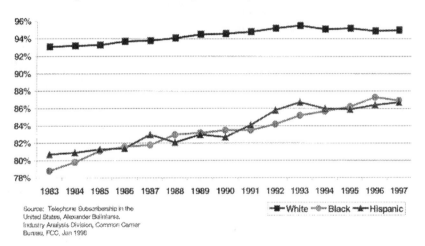

Source: Telephone Subscribership in the United States, Alexander Belinfante, Industry Analysis Division, Common Carrier Bureau, FCC, Jan 1998

White   Black   Hispanic

GEOGRAPHY AND THE TELEPHONE GAP

A review of state data on total telephone penetration rates makes evident that penetration rates vary dramatically from state to state. At one end stand Nebraska and Pennsylvania with all-time highs of 97.1%, and at the other end of the range is New Mexico with a telephone penetration rate of 88.1%.[19] The variation in the experiences of the

18. *Housing of American Indians on reservations—Equipment and fuels.* (1995). Washington, DC: U.S. Bureau of the Census.

19. Table 2, Belinfante, A. (1998). *Telephone subscribership in the United States.* Washington, DC: Industry Analysis Division, Common Carrier Bureau, Federal Communications Commission. Released January 1998.

states leads us to suggest that geography might also influence telephone penetration.

However, taking this approach poses problems. We have as yet not been able to access data that allow us to correlate ethnicity and telephone penetration by geographic unit. That is, data on telephone penetration by state are available, but few data are available that tabulate telephone penetration by ethnicity by state, county, or even census block. What information we have found is not encouraging; we are still looking.

An examination of three states with relatively high penetration rates—California (94.3%), New York (94.2%), and Pennsylvania (97.1%)—illustrates the potential importance of geography as an explanation for ethnicity in the telephone gap.[20]

In California counties for which data are available, the differences between Black (labeled *Blacks* in the Census) and Whites vary in the extreme. Yuba County has a gap of 37.24%, whereas nearby Sacramento County shows almost no difference. Furthermore, five of the counties measured indicate higher telephone penetration rates among Black households.

New York exhibits a range similar to California's. In New York's upstate Jefferson County, White households with telephones outscore Black households by 41 percentage points, whereas in Genesee County the difference between the two groups barely exceeds 2%.

Much the same can be observed in Pennsylvania. York County in the southern tier of the state shows a spread of 12.5 percentage points, Blacks lagging behind Whites. However, Mercer County in the state's northern tier shows Black households leading White households by 1.21%.

Geography may also confound the influence of income. Policymakers and researchers generally consider income as the determining factor in telephone penetration levels. However, there may be a closer relationship between telephone penetration and median household incomes for states with lower median household incomes than in states with greater affluence. If the 10 states with the highest telephone penetration are compared with the 10 states with the highest household median income, only 3 will be found on both lists (Table 11.1). Yet if the lowest 10 states in telephone penetration are compared with the lowest 10 states in household median income, 5 will be found on both lists (Table 11.2). Poorer states do have a tendency to have lower levels of telephone penetration, but the relationship is still somewhat ambiguous; the middle states have a nearly spurious correlation between income and telephone penetration.

20. Table 2, Belinfante, A. (1998). *Telephone subscribership in the United States.* Washington, DC: Industry Analysis Division, Common Carrier Bureau, Federal Communications Commission. Released January 1998.

TABLE 11.1
Ten Highest States for Telephone Penetration and
Median Income Levels in the 1990s

State	State telephone penetration levels 1997	State	State household median income levels, 1995
Pennsylvania	97%	Alaska	$47,954
Nebraska	97%	New Jersey	$43,924
Minnesota	97%	Hawaii	$42,851
Utah	97%	Maryland	$41,041
Iowa	97%	**Wisconsin**	**$40,995**
New Hampshire	**97%**	**Colorado**	**$40,706**
Wisconsin	**96%**	Connecticut	$40,243
Maine	96%	**New Hampshire**	**$39,171**
Colorado	**96%**	Massachusetts	$38,574
Washington	96%	Illinois	$38,071

TABLE 11.2
Ten Lowest States for Telephone Penetration and
Median Income Levels in the 1990s

State	State telephone penetration levels 1997	State	State household median income levels, 1995
Illinois	92%	South Carolina	$29,071
Georgia	92%	Tennessee	$29,015
Arizona	92%	**Louisiana**	**$27,949**
Oklahoma	**91%**	Montana	$27,757
Texas	91%	**Mississippi**	**$26,538**
Louisiana	**91%**	**Oklahoma**	**$26,311**
Washington, DC	91%	**New Mexico**	**$25,991**
Arkansas	**90%**	Alabama	$25,991
Mississippi	**89%**	**Arkansas**	**$25,814**
New Mexico	**88%**	West Virginia	$24,880

Compiled from Table 2, Belinfante, Alexander (1998) *Telephone Subscribership in the United States.* Industry Analysis Division, Common Carrier Bureau, Federal Communications Commission. Released: January 1998; *Statistical Abstract of the United States, 1997: Income and Wealth Section.* Washington, DC: U.S. Bureau of the Census.

What the element of geography seems to say is that the persistence of the telephone penetration gap between the majority and minorities at the national level gives way to significant variations when one focuses down to the level of the counties. Indeed, the county data are important because they illustrate variations in the gap even within the same state, presumably with its uniform public utility commission regulations. In other words, the conditions of the gap seem to vary from state to state and within states, thereby pointing to influences that are local and possibly cultural.[21]

TABLE 11.3
Ten Highest States for PC Penetration and
Median Income Levels in the 1990s

State	State PC penetration levels 1997	State	State household median income levels, 1995
Alaska	55%	Alaska	$47,954
Utah	53%	New Jersey	$43,924
Colorado	52%	Hawaii	$42,851
New Hampshire	50%	Maryland	$41,041
Vermont	47%	Wisconsin	$40,995
Washington	46%	Colorado	$40,706
Idaho	44%	Connecticut	$40,243
Maryland	44%	New Hampshire	$39,171
Virginia	44%	Massachusetts	$38,574
California	43%	Illinois	$38,071

OTHER GAPS

It is true that people are flocking to personal computers and Internet service, but current research indicates that ethnic minorities have yet to reach the penetration levels attained by Whites; the gap may even have grown in the last few years. At present, Whites are twice as likely to own a personal computer as are Blacks or Hispanics—40% compared to approximately 20%. Because personal computers are a newer technology that have not yet had the time to "diffuse" into households, it is not surprising that this penetration gap is more pronounced than the overall gap in telephone penetration (Tables 11.3 and 11.4). Yet between the first and second *Falling Through the Net* reports (1995 and 1998), the gap between Whites and Blacks grew to 21.5%, up nearly 5%; the White–Hispanic gap is now at 21.4%, up nearly 7%.[22] Online penetration rates seem to follow a similar pattern: About 20% of White households are online, whereas about 10% of Black and Hispanic households are connected to the Internet.

---

21. We have recently hypothesized that mobile home ownership among Whites might contribute to the telephone gap because it would award Whites an advantage in demonstrating assets even though their income might be low. However, when we examined California's Los Angeles, Orange, San Bernardino, and Riverside counties—among the foremost in the nation for mobile home ownership—we encountered little or no differences between Whites and African-Americans. Without new evidence, we consider this hypothesis falsified.

TABLE 11.4
Ten Lowest States for PC Penetration and
Median Income Levels in the 1990s

State	State PC penetration levels 1997	State	State household median income levels, 1995
New York	32%	South Carolina	$29,071
South Carolina	31%	Tennessee	$29,015
North Carolina	30%	Louisiana	$27,949
Kentucky	30%	Montana	$27,757
Oklahoma	30%	Mississippi	$26,538
Alabama	29%	Oklahoma	$26,311
Louisiana	25%	New Mexico	$25,991
Arkansas	24%	Alabama	$25,991
West Virginia	24%	Arkansas	$25,814
Mississippi	21%	West Virginia	$24,880

Compiled from Table 2, Belinfante, Alexander (1998) *Telephone Subscribership in the United States.* Industry Analysis Division, Common Carrier Bureau, Federal Communications Commission. Released: January 1998; *Statistical Abstract of the United States, 1997: Income and Wealth Section.* Washington, DC: U.S. Bureau of the Census.

A recent survey conducted by the Tomas Rivera Policy Institute of the Claremont Colleges in California concentrated primarily on the low levels of personal computer penetration in Hispanic households.[23] Although finding that 31% of all Hispanics nationally have yet to use a computer, its overall results were far more encouraging: Hispanics had an online penetration rate of about 30%, continued to outpace all other ethnic groups in acquiring new users, and had doubled the number of computers in their homes in the last 4 years. In addition, Hispanics are joining online services faster than the national average: Between 1994 and 1998, the annual growth rate for new Hispanic online members was 130% compared to the national average of 65%.

This survey also highlighted the importance of education and income. Approximately 77% of Hispanic households connected to the Internet also had some college education, compared to 15% for unconnected households. Over 63% of Hispanic households with incomes over $75,000 are connected to the Internet. Moreover, this tendency appears to parallel the role of income in creating temporary gaps during the diffusion of television.

In fact, median income levels of online users are significantly above the national average. Between 50% and 70% of online users have a median household income

22. *Falling through the Net II: New data on the digital divide.* (1998). Washington, DC: National Telecommunications and Information Administration. Available http:// www.ntia.doc.gov/ntiahome/net2/falling.html, Chart 12, 14, 15.

23. *Closing the digital divide.* (1998). Tomas Rivera Policy Institute.

over $50,000. The average median income level for online users is about $60,000.[24] These numbers indicate that personal computers and Internet services are still largely the domain of the affluent, but decreasing computer ownership costs and stable one-price monthly Internet connection fees will likely encourage diffusion of these two technologies across income ranges.[25]

Because education interacts with income in affecting computer purchases, minorities with historically lower rates of education and income are likely to lag in their adoption of personal computers. Still, the example of Hispanics and computer adoption points to a parallel with the fleeting gap in TV diffusion rather than the persistent gap in telephone penetration. Computer purchases are likely to be led by wealthier, educated households, followed by less wealthy but educationally oriented households. Thus, we suggest that the "personal computer gap" is temporary as was the "TV gap." What remains unclear is the rate at which computers will be adopted and the final saturation level in household penetration.

SOME CONCLUSIONS AND SPECULATIONS

- Ethnicity appears to be a characteristic in most of the gaps identified in this chapter. It appears over and over, yet does not stand out clearly and apart from other identified factors.
- The factors seemingly present and contributing to the gaps identified here intermix in ways that make parsing them out nearly impossible given the statistical data available. For this reason, it seems highly probable that an individual is affected by multiple factors that may converge and tip the scales against telephone subscribership. Minorities may well be inordinately affected by multiple factors.
- Not all gaps persist. In the case of the radio and the television, whatever gaps existed during the early period of diffusion closed with the approach of saturation. What ties these two media together is their exchange in the marketplace as goods with a simple purchase structure. There are some indications that the same may hold for computers.

---

24. Source: *Pulse—Consumer Profiles: Changing in a High Tech World,* 5(3), June 1998, CTAM; *Closing the digital divide,* Tomas Rivera Policy Institute, 1998; MediaMark Cyber-Stats, Spring 1998 at www.mediamark.com; NetRatings, Inc. news release December 8, 1997, available www.netratings.com/newsDec_8.htm.

25. For instance, few computers sold in the retail market cost less than $1,000 in 1996. Yet by 1997, sub-$1,000 computers constituted 48.4% of retail computer sales, up from 1.2% in 1996. Computers more than $2,000 now constitute less than 8% of retail computer sales. J. Kirchner, PC prices: How long can they go?, PC Magazine Online, March 10, 1998, available http://www.zdnet.com/pcmag/issues/1705/283015.htm; Wasserman, T. (1998). Expensive PCs face challenge, *Computer Retail Week* using International Data Corporation (IDC) data, June 15, 1998.

- Geography appears to influence the existence of the telephone gap. Even within a state, large variations emerge across counties. This finding indicates that local circumstances may be the most important contributors to the existence of gaps, the telephone gap in particular.

The existence of gaps—the telephone gap in particular—should trouble anyone concerned for the future of participatory democracy in the information age. Full-fledged access in the 21st century means at least a telephone in the home and a computer with Internet access. Those without telephones lack effective participation in rapidly expanding digital Internet communities. If historical telephone penetration gaps cross over into the personal computer and Internet user populations, it is quite possible that a persistent and ethnicity-centered gap between the information haves and have nots will emerge, much to the detriment of an integrated society. Ironically, the telephone has rapidly become the weak link in the IT chain: It is now a necessary but not sufficient technology in the modern home.

If we wish to solve the mystery of the gaps, we will have to look beyond the data that have guided us in the past. We must go beyond national data to data specifically describing conditions at the levels of the states, counties, and perhaps even census blocks. When we do this, we are likely to find a complex array of factors more particular to localities than to the country as a whole. All the while, we should remind ourselves to avoid thinking of all gaps as sharing the same characteristics.

For Langston Hughes's lamentation to become history, we will have to dedicate ourselves more diligently to probing the meaning of these gaps and to solving them. Thus, the challenges raised by Larry Irving and William Kennard ring loudly because the consequences stand starkly on the horizon.

ACKNOWLEDGMENTS

This chapter was made possible by a gift from Bell Atlantic to the Institute for Information Policy in support of research on universal service. We are especially indebted to Ed Lowry of Bell Atlantic for his advice and support. The chapter reflects only the views of the authors.

Do Commercial ISPs Provide Universal Access?

Thomas A. Downes
Tufts University

Shane M. Greenstein
Northwestern University

Concern over the potential need to redefine universal service to account for Internet-related services and other combinations of communication and computing motivates this study of the geographic spread of commercial Internet service providers (ISPs), the leading suppliers of Internet access in the United States. We characterize the location of 54,000 points of presence (POPs), local phone numbers offered by commercial ISPs, in the spring of 1998. Markets differ widely in their structure, from competitive to unserved. More than 92% of the U.S. population has easy access to a competitive commercial Internet access market, while less than 5% of the U.S. population has costly access.

MOTIVATION

Governments frequently revisit the principle of universal service, making it an enduring issue in communications policy. In past eras this goal motivated policies that extended the national telephone network into rural and low-income areas. In the last few decades the same concerns motivated policies that eliminated large disparities in the rate of adoption of digital communication technology within the public-switched telephone network. More recently, many analysts have begun to anticipate a need to redefine universal service to account for Internet-related services and other combinations of communication and computing.[1]

The U.S. population is far from universal adoption of Internet services at home. No survey shows more than 25% adoption of Internet access in U.S. households, and no survey shows personal computer adoption exceeding 45% of households.[2] Yet, these surveys all beg the question about the availability of access—in other words, whether all households have access to Internet service at the same low cost. Because all consumers have access to the Internet at some price, the key question for most consumers is whether they can "cheaply" access the Internet. For many users *cheap* is synonymous with a local telephone call to a firm that provides Internet access for a fee. This open question motivates this study, which documents and analyzes the geographic spread of commercial Internet access.

This study focuses on understanding the geographic spread of commercial ISPs, the leading suppliers of Internet access in the United States. In the absence of changes in government policy, market-based transactions with ISPs will be the dominant form for delivery of online access for medium and small users.

Although our focus is the geographic spread of ISPs, our study will influence many facets of the literature on information infrastructure policy and cybergeography.[3] Because ISPs use the public-switch network, policy for this network should be sensitive to the commercial forces in the ISP market. There is no reason to anticipate that commercial ISPs provide universal access. Indeed, since there is little research into the organization of this industry, there is almost no framework to use for speculating.

In the first part of this chapter, we provide a brief introduction to the geographic diffusion of commercially oriented Internet access providers.[4] In the second part of the chapter we address a related empirical issue: Do all regions of the country receive similar access to Internet services provided by commercial firms? To answer this question, we characterize the location of more than 54,000 dial-up access points offered by commercial ISPs in Spring 1998.

Spring 1998 was a good time for such a survey. The industry's structure, although not completely stable, was not changing drastically every month. Most

1. For example, "The traditional concept of universal service must be redefined to encompass a concept more in line with the information superhighway of the future" (Anstey, 1993). That policymakers are sensitive to such statements is apparent in the 1996 Telecommunications Act, which contains provisions to collect funds to finance the diffusion of Internet access to public institutions, such as schools and libraries.

2. See, for example, Kridel, Rappaport, and Taylor (1997), Maloff Group International (1997), Compaine and Weinraub (1997), or Clemente (1998).

3. A complete bibliography is impossible. For some recent studies, see the references in Downes and Greenstein (1998).

4. The conclusions from the first half of the chapter will be familiar to regular readers of commercial press for the ISP industry. For surveys of the online industry and attempts to analyze its commercial potential, see Meeker and Dupuy (1996), Hoovers Business Press (1997), Juliussen and Juliussen (1996), or Maloff Group International (1997).

firms had been in the ISP market for a few years, making it possible to document their strategies, behavior, and commercial achievement. The key findings of the empirical work are as follows:

- The U.S. commercial ISP market is comprised of thousands of small geographically dispersed local markets for Internet access. There is no single structure that characterizes the ISP market across the country, and we should not expect this heterogeneity to disappear.
- More than 92% of the U.S. population had access to a competitive local ISP market. Less than 5% of the U.S. population lived in areas with no access to any provider and approximately 3% lived on the margin between easy access and no access.

THE STRATEGY AND ORGANIZATION OF THE ISP BUSINESS

This study examines the Internet access business just over 3 years after the National Science Foundation relinquished rights over the Internet to commercial entities. By this time many firms understood the technology for the delivery of Internet access using TCP/IP, but the commercial norms for the business were in flux. Different organizations employed different commercial models for the delivery of Internet access. During this experimentation, Internet access spread to many different regions of the United States.

Scope of Investigation

We analyze the ISP industry after the development of browsers and therefore focus attention on firms that provide dial-up service that enables a user to employ a browser. Furthermore, we make no distinction between firms that began as online information providers, computer companies, telecommunications carriers, or entrepreneurial ventures. As long as their ultimate focus is commercial Internet access either as a backbone or a downstream provider, they will all be characterized as ISPs.

Access and Location

There is a growing market for direct access through the use of competitive access providers, but this service is primarily focused on business use within big cities. Many of the ISPs we analyze also provide direct access to business in the areas they serve (e.g., by building, renting, and maintaining T1 lines); but this activity will be in the background, as this study focuses on the dial-up market. There is also a market for 800 dial-up access, which, because of its expense for heavy usage, is targeted at business users who have occasional high-value online needs away from home (Barrett, 1997; Boardwatch Magazine, 1998; Maloff, 1997).

Most universal access issues concern the adoption rates of medium and small users, since these are the users on the margin between no access and a few low-cost alternatives. ISPs targeting users with regular and modest needs, which describes most residential users and small businesses in the United States, require the user to make phone calls to a local switch. The cost of this call depends on mostly state regulations defining the local calling area and both state and federal regulations defining the costs of long-distance calling. The presence of ISPs within a local call area, therefore, determines a user's access to cheap Internet service. Similarly, the number of local ISPs determines the density of supply of low-cost access to Internet services within a small geographic region. Thus, the geographic spread of ISPs determines the cost of Internet access for most of the marginal users of the Internet.

The Maturing of the ISP Industry

As recently as 1995, only a few enterprises offered national dial-up networks with Internet access (Boardwatch Magazine, 1998), mostly targeting the major urban areas. At this time it was possible to run a small ISP on a shoestring in either an urban or rural area. In contrast, by Spring 1998, there were dozens of well-known national networks and scores of lesser known national providers. There were also many local providers of Internet access that served as the links between end-users and the Internet backbone. Shoestring operations seemed less common.

As a technical matter there is no mystery to starting and operating a dial-up ISP. A bare-bones independent ISP requires a modem farm, one or more servers to handle registering and other traffic functions, and a connection to the Internet backbone.[5] As an economic matter, starting and operating a node for a dial-up ISP involves many strategic considerations (Maloff, 1997; Stapleton, 1997). Higher quality components cost money and may not be necessary for some customers. High-speed connections to the backbone are expensive, as are fast modems. Facilities need to be monitored, either remotely or in person. Additional services, such as Web hosting and network maintenance for businesses, are also quite costly, as they must be properly assembled, maintained, and marketed. Providing added value may, however, be essential for retaining or attracting a customer base.

In sum, the geographic reach and coverage of an ISP is one of several important dimensions of firm strategy. Geographic coverage is determined in conjunction with choices of value-added services, scale, performance, and price. Providers that seek to provide national service must choose the regions in which they maintain POPs. The commercial motives of providers would lead us to expect to find that

---

5. For example, see the description in Kalakota and Whinston (1996), Leida (1997), the accumulated discussion on www.amazing.com/internet/faq.txt, or Kolstad's (1998) remarks at www.bsdi.com.

national firms cover areas of the United States that contain most of the population. In addition, local ISPs would be expected to target many of the niche markets that the national ISPs fail to address, especially those niches where users require a "local" component or customized service.

EMPIRICAL RESEARCH QUESTIONS

On the basis of the foregoing reasoning, we predict that most urban areas will have abundant Internet access from commercial firms and some remote areas might not. Between these two predictions lies a very large set of possibilities. Narrowing this set of possibilities is the focus of the following empirical work.

- *Question 1—The extent of geographic coverage*: Some parts of the country will not have access to low-cost commercial Internet providers. How does access change when density increases? What conditions characterize the competitive areas?

- *Question 2—The degree of competition in urban and rural areas:* We expect most residents of urban and high-density areas will face a competitive and abundant supply of Internet access from commercial firms. What fraction of the population living in such areas has access to competitive ISP markets?

DATA

To track the geographic spread of ISPs, we compiled a list of telephone numbers for dial-up access and their location. We then computed the geographic distribution of the POPs across the United States. We explain these data and methods in the following.

Data Sources

The best way to compile a list of ISPs by location is to go to the information sources used by most potential ISP consumers. Although there is no single "yellow pages" for ISPs, there are a few enterprises that track ISPs in the United States. In Spring 1998, we surveyed every compilation of ISPs on the Internet. This study's data combine a count of the ISP dial-in list from Spring 1998 in *thedirectory* and a count of the backbone dial-in list for Spring 1998 in *Boardwatch* magazine.[6]

This choice was made for several reasons. First, *thedirectory* requests that its ISPs list the location of their dial-in phone numbers. Although not all of the ISPs

---

6. Current versions of these lists may be examined at www.thedirectory.org and www.boardwatch.com. This includes POPs found in the ISP section of *thedirectory*. This also includes POPs for ISPs listed in the *Boardwatch* backbone section.

comply with this request, most do, making it much easier to determine an ISP's location in a general sense. Second, *thedirectory* and *Boardwatch* both claim to maintain comprehensive lists, and these claims seem to be consistent with observation. That said, *thedirectory* consistently lists more ISPs than *Boardwatch*. On close inspection, it appears that this results from *thedirectory*'s more extensive coverage of small ISPs. Third, whereas *thedirectory* shows the location of most ISPs, *Boardwatch* only does so for backbone providers. In sum, *thedirectory* ISP list contains a more comprehensive cataloging of the locations of POPs maintained by all ISPs except the national backbone providers, for which *Boardwatch* contains a superior survey of locations.

We used the following strategy to determine the location of each POP: When the city of a dial-in phone number was listed, we assumed the POP was in that city. When it was in doubt, the area code and prefix of the dial-in POP were compared to lists of the locations of local switches with these area codes and prefixes. We used the location of the local switch in that case. If this failed to locate the POP, which happened for small ISPs that only provided information about their office, then we went to the Web page for the company. If there was no information about the company's network, the voice dial-in number for the ISP headquarters was used as an indicator of location.[7] When a city overlapped two counties and the phone number could not be used to place the POP in a county, the POP location was assumed to be the county in which the city had the greatest share of its land.

On final count, *thedirectory* contained 49,472 POPs not found in *Boardwatch*, *Boardwatch* contained 3,627 POPs in its backbone list not found in *thedirectory*, and 1,360 phone numbers came from both. The merged set contained 54,459 POPs that served as dial-in POPs. These phone numbers corresponded to a total of 41,117 unique firm-county presences, because many firms maintained multiple POPs in the same county, for 6,006 ISPs. Of these firm-county presences, more than three fourths were associated with just over 200 firms. Of the total number of ISPs, approximately half were ISPs for which we had only a single phone number.

Any conclusions reached here are potentially invalid if the data construction procedure generated sampling error that correlates with geography. We think not, although the preceding procedures may have imparted some small biases to some counties, which we describe later. Overall, there appears to be no strong evidence of any error in the coverage of small commercial ISPs. In addition, there appears to be a strong positive correlation in the geographic coverage of national firms be-

---

7. This last procedure mostly resulted in an increase in the number of firms we cover, but not a substantial change in the geographic scope of the coverage of ISPs. The data set contained 45,983 phone numbers prior to adding firms that only provided a Web address. The additional 1,348 ISPs were responsible for 8,476 firm-county presences, which did not disproportionately show up in uncovered areas. They did, however, help identify entry of ISPs in a few small rural areas.

cause most of them locate predominantly in urban areas. Thus, even if these two lists failed to completely describe the coverage of many national firms, it is unlikely that the following qualitative conclusions would change much if the omitted POPs were added.[8]

The preceding procedures may show less ISP entry than has actually occurred in counties that border on dense, urban counties. There is a tendency for new suburbs to use the telephone exchange of existing cities, which may be just over the county border. Unless the ISP specifically names this new suburb in the bordering county as a targeted area, our procedures will not count the ISP's presence in that new suburb.

A similar and related bias arises when a county's boundaries and a city's boundaries are roughly equivalent, even when the neighboring county contains part of the suburbs of the city. In this situation, many ISPs will claim to be located within the city's boundary even though residents will recognize that the ISP is located on the city boundary and the coverage of the ISP may be more extensive than this declaration would indicate. We control for these potential biases through tables that treat as the market a county and its nearby neighbors.

In the best case scenario, the compilation in this study will give an accurate account of all commercial ISP coverage in the United States, particularly the coverage of those companies that advertise through standard channels. In the worst case scenario, counting the locations of the POPs listed in both directories will give an indication of how the ISP market looks to a consumer who does a small amount of searching. The compilation in this study probably lies between the worst and best cases, both of which are acceptable for a study of the spread of ISPs across the nation.

Definitions

What types of ISPs are on these lists and does their selection have any implications for the scope and coverage of this study? Both *thedirectory* and *Boardwatch* try to distinguish between bulletin boards and ISPs, where the former may consist of a server and modems, whereas the latter provide Web access, file transfer protocol, e-mail, and often much more.[9] Thus, the scope of this study is appropriately limited to firms providing commercial Internet access. We also excluded firms that only provided direct access.

---

8. Indeed, we tested this proposition on the data in the study. Even if a dozen national providers were left out of the sample, the basic qualitative conclusions would not change. For further evidence on the difference between the geographic coverage of local, regional, and national firms, see Downes and Greenstein (1998).

9. Extensive double-checking verified that *thedirectory* and *Boardwatch* were careful about the distinction between an ISP and a bulletin board.

Second, both lists concentrate on the for-profit commercial sector. For example, both eschew listing university enterprises that effectively act as ISPs for students and faculty. This is less worrisome than it seems, as commercial ISPs also gravitate toward the same locations as universities. This study's procedure, therefore, picks up the presence of ISP access at remotely situated educational institutions unless the amount of traffic outside the university is too small to induce commercial entry.

Thedirectory does list some free-nets. Their inclusion appears to depend on whether the free-net notifies *thedirectory* of their existence. A similar remark can be made for local cooperatives and quasi-public networks that are part of a state's educational or library system. In general, this study's procedures will identify the commercialized sector of Internet access but may underrepresent some nonprofit access alternatives, especially those that do not advertise in the standard online forums.

The tables provided later give a broad description of county features. Population numbers come from 1997 U.S. Bureau of the Census estimates. We label a county as urban when the Census Bureau gives it a metropolitan statistical area (MSA) designation, which is the broadest indicator of an urban settlement and includes about a quarter of the counties. For all tables, the data pertain to all states in the continental United States. These data also include the District of Columbia, which is treated as another county. County definitions correspond to standard Census Bureau county definitions, resulting in a total of 3,110 counties.

It is well known that slicing U.S. geography in this way has certain drawbacks— principally, county boundaries are political boundaries and do not directly correspond with meaningful economic market boundaries. We think that these drawbacks are overwhelmed by the benefits of using Census Bureau information. Moreover, it is also possible to control for the worst features of this drawback by calculating statistics that account for nearby counties.[10] Each of the 3,110 counties is the elemental observation, but in calculating some of the summary statistics we use as the unit of observation the county together with some of the nearby neighboring counties. We define nearby counties as counties with a geographic center, as defined by the Census Bureau, within 30 miles of the geographic center of the county of residence. We chose 30 miles because this is within the first mileage band for a long-distance call in most rural areas.[11]

---

10. To see it both ways, see Downes and Greenstein (1998). We make this calculation using the U.S. Bureau of the Cens2us's (1992) *Contiguous County File, 1991: United States*.

11. We experimented with a number of different mileage bands. We tried 15 miles and found that the results were qualitatively no different from using no information about county neighbors. We also tried all neighboring counties without distinguishing by their distance and also found that this was far too inconclusive, especially in the western United States. These latter results are included in the appendix of Downes and Greenstein (1998).

Maps

Figure 12.1 illustrates the density of location of ISPs across the United States at the county level. Black and gray areas indicate the extent of entry. White areas have none. The picture illustrates the uneven geographic coverage of the ISP industry. ISPs tend to locate in all the major population centers, but there is also plenty of entry into rural areas. The map also illustrates the importance of accounting for the influence of nearby counties.

FIGURE 12.1
Distribution of ISPs: May 1998.

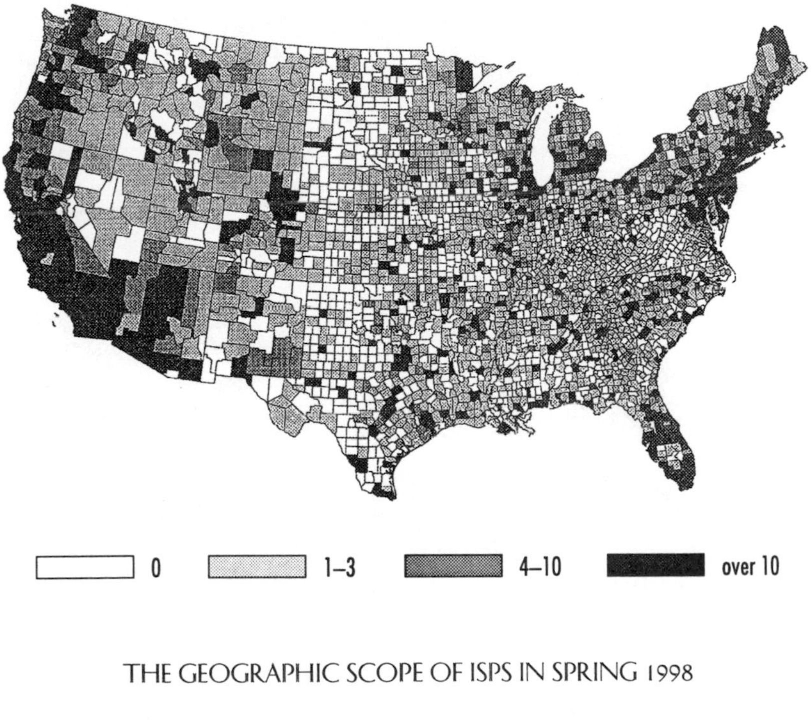

| ☐ 0 | � 1–3 | ▇ 4–10 | ▇ over 10 |

THE GEOGRAPHIC SCOPE OF ISPS IN SPRING 1998

The summary of the nature of ISP coverage can be found in Tables 12.1, 12.2, and 12.3.

Urban–Rural Differences

Table 12.1 is organized by counties in the continental United States, where the central county is the unit of observation. Of the 3,110 counties, 247 have no POP supported by an ISP within their boundaries or the boundaries of a nearby county,

141 have only one, 191 have only two, and 115 have only three. Just under 4.5% of the U.S. population lives in counties with four or fewer ISPs nearby. As further evidence that low (high) entry is predominantly a rural (urban) phenomenon, more than 95% of the counties with 10 or fewer suppliers are rural.

TABLE 12.1
Entry of ISPs: Number of Providers in the "Market"
Market Definition: County of Residence and All Counties Within 30 Miles

Number of providers in market	Number of counties	% of population in these counties	Cumulative population %	% of these counties urban
0	247	0.84	100.00	1.21
1	141	0.59	99.16	4.96
2	193	1.12	98.57	4.15
3	115	0.69	97.45	6.09
4	172	1.23	96.76	5.23
5	98	0.81	95.53	6.12
6	117	1.10	94.71	5.13
7	81	0.71	93.62	3.70
8	83	1.00	92.91	13.25
9	73	0.77	91.91	4.11
10	72	0.71	91.14	2.78
11–15	278	3.06	90.42	7.91
16–20	147	1.88	87.37	14.97
21 or more	1,293	85.48	85.48	56.46

Note: For the calculations in the final three columns, the county of residence is treated as the unit of observation.

Table 12.1 offers the starkest finding of this study. More than 92% of the U.S. population has access by a short local phone call to seven or more ISPs. Moreover, the geography of the universal access issue, as of Spring 1998, was predominantly rural and, then, only pertinent to a fraction of the rural population.

Table 12.2 elaborates on Table 12.1, giving the relationship between the presence of ISPs and some basic features of counties, principally population and population density. Although there is variance around the relationship between population and the presence of ISPs, the trend in average population size is almost monotonic. That is, for small markets the number of suppliers grows with population. This result holds whether the market definition accounts for neighboring counties or does not. Density is also correlated with entry, a result that is more apparent when one accounts for neighboring counties.

Table 12.3 permits exploration of the difference between urban and rural areas, shedding light on the relationship of density to entry. Counties are divided into those in urban and rural areas, and for each division the same summary statistics as in Table 12.2 are provided. Apparent is the difference between urban and rural areas in the relationship between the population and density of the region and ISP entry. In rural areas, both population levels and density predict the level of entry into small counties (those with fewer than five suppliers). In contrast, in urban areas population levels do not strongly coincide with ISP entry. Because there are, however, few urban area observations with 10 or fewer ISPs, these results reflect a few observations and do not indicate a broad trend.

TABLE 12.2
Population and Population Density by Number of Providers in the "Market" Market Definition: County of Residence and All Counties Within 30 Miles

Number of providers in market	Number of counties	Mean population: county	Mean population: market	Mean population density: county	Mean population density: market
0	247	8,971.00	27,441.41	8.75	9.25
1	141	10,964.23	54,761.48	15.84	18.04
2	193	15,352.19	57,342.54	16.18	17.26
3	115	15,788.00	79,514.72	23.34	25.24
4	172	18,879.13	78,464.70	30.84	25.86
5	98	21,886.73	121,563.11	35.11	39.03
6	117	24,705.74	104,953.21	31.83	33.50
7	81	23,107.25	144,111.94	84.29	46.90
8	83	31,646.20	141,177.76	40.89	44.36
9	73	27,853.75	175,581.14	60.81	56.47
10	72	26,143.24	131,086.67	40.52	41.37
11–15	278	28,975.35	186,185.63	55.59	61.97
16–20	147	33,781.75	224,800.27	75.80	73.40
21 or more	1,293	174,195.64	943,724.38	506.30	327.82

Note: Population density is measured as population per square mile.

Finally, the results in Table 12.3 are consistent with the view that there are economies of scale at the POP and that these economies largely determine the relationship between number of suppliers and population levels in rural areas. If there are economies of scale at the POP and no difference in demand across regions with different density, then economies of scale determine threshold entry and incremental entry thereafter. Table 12.3 does not provide conclusive evidence of these scale economies, however, because in constructing this table we did not control for potential determinants of demand. It is possible that different geographic features of these areas may correlate with different levels of demand or unobserved intensities of demand that systematically differ across counties of different population size. There is, however, insufficient evidence in these tables to test these competing hypotheses.

TABLE 12.3
Population and Population Density by Number of Providers
in the "Market" and by Urban–Rural Status
Market Definition: County of Residence and All Counties Within 30 Miles

Number of providers in market	Number of counties	Mean population: county	Mean population: market	Mean population density: county	Mean population density: market
Rural counties					
0	244	8,433.93	26,437.52	7.86	8.42
1	134	8,737.10	48,970.83	13.24	16.14
2	185	13,589.57	52,514.58	14.27	15.52
3	108	13,622.34	73,110.43	20.45	22.90
4	163	18,228.04	75,663.53	29.48	24.47
5	92	19,560.42	113,722.88	32.75	36.62
6	111	22,875.61	100,510.03	28.90	31.84
7	78	22,789.35	137,283.77	83.95	44.53
8	72	23,164.31	125,816.53	32.28	38.37
9	70	26,247.36	160,698.86	58.11	49.38
10	70	25,083.89	129,346.40	39.56	40.71
11–15	256	25,579.48	172,627.71	49.84	56.59
16–20	125	29,560.07	201,130.22	71.05	63.01
21 or more	563	37,470.21	331,932.03	91.14	105.12
Urban counties					
0	3	52,652.67	109,091.33	80.83	76.70
1	7	53,597.71	165,611.00	65.66	54.38
2	8	56,112.75	168,989.13	60.31	57.45
3	7	49,201.00	178,323.86	68.00	61.44
4	9	30,671.00	129,196.89	55.44	51.17
5	6	57,556.83	241,780.00	71.26	76.05
6	6	58,563.00	187,152.00	85.91	64.16
7	3	31,372.67	321,644.33	93.14	108.65
8	11	87,164.09	241,724.0	97.27	83.54
9	3	65,336.33	522,834.33	123.69	221.93
10	2	63,220.50	191,996.00	73.96	64.49
11–15	22	68,490.86	343,950.50	122.45	124.54
16–20	22	57,768.55	359,289.18	102.76	132.41
21 or more	730	279,642.79	141,558.75	826.48	499.58

Note: Population density is measured as population per square mile.

In the absence of some decreasing cost technology, such as some sort of coordination economies, or increasing returns on the demand side, these basic economics limit the geographic expansion of the national ISP networks. Because national firms face constant costs to the addition of POPs, they will not expand their network POPs in increasingly remote areas, bringing in fewer additional customers with each additional expansion. Hence, no national firm finds it economic to be ubiquitous.

Implications for Geographic Scope of Commercialized Internet Access

Several findings from these tables should shape further policy discussions of the commercialization of Internet access. First, the diffusion of Internet access is a commercial process driven by commercial motives. Nevertheless, the firms in this industry have developed Internet access markets for most of the U.S. population in a relatively short period. Second, some regions of the country, primarily less densely populated rural areas, do not have access to any low-cost commercial Internet providers. There is a minimum threshold of population needed to support entry of an ISP POP, although local and national POPs may face different thresholds. Third, some regions face competitive access markets and some do not. Most residents of urban and high-density areas have a competitive and abundant supply of Internet access from commercial firms. The part of the U.S. population that does not have access to a competitive ISP market lives in rural and low-density areas. We develop a few additional implications in the following.

THE SCALE OF ISPs AND SCOPE OF GEOGRAPHIC COVERAGE

Downes and Greenstein (1998) showed that local and national firms place POPs at different locations, with national firms avoiding small, less densely populated rural areas. This pattern is consistent with three theories. First, local POPs in rural areas may enter with lower quality than national POPs. That is, entering with low-quality equipment lowers a local POP's costs. Alternatively, local POPs in rural areas may be entering with different value-added services than national POPs in urban areas. That is, local POPs in rural areas may not be deriving much profit from their ISP service, but they make up for these losses with other complementary services tailored to rural areas. The second view only makes sense if the value-added services offered by a local POP have a strong local component; otherwise, a national firm could imitate it and profitably expand into rural areas. A third view[12] is that many rural ISPs provide service as part of their activities as rural cooperatives or other quasi-public institutions supporting local growth. In this view the desire to provide community and public service, and not the profit motive, is the key driver of entry in rural areas. This different motive would account for the willingness of rural ISPs to enter areas that profit-oriented, national ISPs avoid. It is still largely a matter of speculation about which view is most likely.

---

12. See, in particular, Garcia (1996) and Garcia and Gorenflo (1997).

These three views will set the agenda for the universal access debate into the next century. If there are strong economies of scale at the POP, these will limit entry of ISPs in rural and remote areas. If ISPs become essential for local growth, there may be a role for public or quasi-public local institutions to subsidize local ISPs to overcome their inability to take advantage of these scale economies. If the local component of an ISP's service becomes an essential element of its offerings, then national firms may never find it commercially profitable to move to remote areas. If high-quality service is expensive to offer, then remotely situated firms in rural areas may find it difficult to afford to upgrade their networks. Of course, all of this could change if scale economies weaken or if the costs between high and low quality narrow enough so that ISP product lines become similar in rural and urban areas.

MARKET STRUCTURE, TAXATION, AND SUBSIDIES

These patterns should influence any debate about subsidies and taxation of the ISP industry. All future policy debates should be cognizant of the fact that changes in policy will affect urban and rural areas differently. For example, altering access charges for ISPs will elicit different responses depending on whether the area is predominantly served by local or national companies. Similarly, taxing ISPs, which many states are already doing or proposing, will produce differences between urban and rural communities. If the percentage of revenue associated with non-dial-up business differs between ISPs in urban and rural areas, the same tax could result in altering the mix of services offered in each type of area.

Proposals for subsidizing ISPs also bring forth some difficult questions. First, the foregoing results make clear that few residents of the United States have no access to the Internet. Thus, universal subsidies to ISPs in urban areas and other competitive markets seem unjustifiable. Second, if private firms stay out of rural areas they do so for sound economic reasons. Only compelling social benefits justify ignoring these reasons. Some critics charge that ISPs are already receiving a large implicit subsidy by not paying for access.[13] Are the social benefits of extending ISPs further subsidies, even if those subsidies are targeted, worth an increase in these social expenses?

COMMERCIAL MOTIVES AND PUBLIC SUPPORT FOR THE INTERNET

In the years leading up to the commercialization of the Internet, government support took many forms. The federal government provided subsidies for access to In-

---

13. For example, contrast the very different proposals in Sidak and Spulber (1998) and in Garcia and Gorenflo (1997). The former call for an end to implicit subsidies, and the latter come close to calling for subsidies for rural ISP service.

ternet protocol networks at remotely situated universities, software and shareware development, the development of backbone infrastructure, and the operation of many governance mechanisms. The commercialization of the Internet and the explosive growth that followed raise new questions about the proper role, if any, for government support in the future. In the spirit of this inquiry, we note that it is tempting to interpret our findings as indicating that commercial firms went far toward meeting goals for universal access. This would seem to lead to the conclusion that commercial firms accomplished much of this without government support. We must properly qualify such a conclusion.

First, in the mid-1990s the commercial Internet access industry still retained significant indirect technical support from university computer science and engineering programs, where federal government research support continued. This support took the form of research grants for the development of software or hardware technologies and subsidies for the training of advanced engineers. This indirect subsidy for the industry as a whole cannot be disentangled from the commercial behavior we observe.

Second, the geographic shape of the commercial Internet access market may still retain significant imprint of federal and state support for the development of information infrastructure. For example, much backbone was laid in the 1980s and 1990s to support traffic flows between universities and government research facilities or to support other educational needs. Many state governments also developed fiber lines in parts of their state in support of similar initiatives. The firms associated with that backbone, such as MCI, are still major providers of backbone today and still use much of the same backbone for commercial traffic. Significant commercial developments engendering new traffic patterns will alter those configurations in due time, but initially these commercial uses were built on top of the old structures. Some part of this infrastructure is attributable to the old subsidies, its expenses are sunk and providing services, and it should be properly called a subsidy today.

Third, providers might have entered in anticipation of future federal support. Anticipated support may take many forms. Recent programs, such as the development of Internet II and the disbursement of universal service funds as mandated by the 1996 Telecommunications Act, have received much publicity. We have no way to tell how much of the observed commercial behavior today can be attributed to investments made in anticipation of these expected funds.

Finally, our study has documented the lowest cost and lowest quality sector of the Internet access industry. Much of this industry exists in conjunction with the provision of other commercial services and is influenced by government support. Existing and future government programs may, directly or indirectly, encourage demand for high-speed access, thus influencing the overall profitability of an ISP business in a particular area. As a by-product, these programs may also lead to the development of the dial-up industry in a local area.

CONCLUSIONS

The commercial Internet access industry has an important geographic component that correlates with features of market structure, quality of service, pricing, and competitiveness. As a result, most of the important issues in the universal access debate have an important geographic component. The links between geographic coverage and market structure arise because an ISP simultaneously chooses several important dimensions of firm strategy, including geographic coverage.

The location pattern we observed in Spring 1998, particularly the failure of ISP service to spread to all parts of the country, is consistent with the existence of small economies of scale at the POP. Related strategic decisions induced variance in market structure in different regions of the country. The end result is that most of the population faces competitive supply of Internet access, although some residents of rural areas faced less ideal conditions.

These structural and strategic differences should be central issues in policy discussions of universal access to advanced communications and computing technology. Many issues will remain unresolved until future research on access analyzes the precise determinants of firm entry and expansion strategies. To what extent is entry influenced by the presence of a wealthy or educated population, an advanced telecommunications infrastructure, or a major educational institution? Answering these questions is a necessary first step toward properly structuring universal access policies.

ACKNOWLEDGMENTS

This study was funded by the Institute for Government and Public Affairs at the University of Illinois and by a Mellon Small Grant in the Economics of Information at the Council on Library Resources. We appreciate comments from Amy Almeida, Tim Bresnahan, Linda Garcia, Zvi Griliches, Padmanabhan Srinagesh, Pablo Spiller, Dan Spulber, Scott Stern, and participants at the NBER productivity lunch seminar, Stanford Computer Industry Project seminar, Northwestern Communications Colloquia, University of Kansas Technology Seminar, Consortia for Telecommunication Policy Research 1998 conference in Ann Arbor, and participants at the Harvard Information Infrastructure Project workshop on "The Impact of the Internet on Communications Policy." Howard Berkson, Heather Radach, Holly Gill, and Helen Connolly provided excellent research assistance at various stages. We would especially like to thank Angelique Augereau for her observations and extraordinary research assistance in the final stages. We take responsibility for all remaining errors.

REFERENCES

Anstey, M. (1993). *A nation of opportunity: Realizing the promise of the information superhighway.* DIANE Publishing Co.

Barrett, R. (1997, March 24). Office buildings link to Internet backbone. *Inter@ctive Magazine.* Available http://www.zdnet.com/intweek/print/970324/inwk0070.html

Boardwatch Magazine (1998). *March/April directory of Internet service providers.* Littleton, CO.

Clemente, P. (1998). *The state of the Net, the new frontier.* New York: McGraw-Hill.

Compaine, B., & Weinraub, M. (1997). Universal access to online services: An examination of the issue. *Telecommunications Policy, 21*(1), 15–33.

Downes, T., & Greenstein, S. (1998). Universal access and local commercial Internet markets [mimeo]. Available http//: skew2.kellogg.nwu.edu/~greenste/

Garcia, D. L. (1996, November–December). Who? What? Where? A look at Internet deployment in rural America. *Rural Telecommunications,* pp. 24–29.

Garcia, D. L., & Gorenflo, N. (1997, September). *Best practices for rural Internet deployment: The implications for universal service policy.* Paper presented at 1997 Telecommunications Policy Research Conference, Alexandria, VA.

Hoovers Business Press. (1997). *Hoover's guide to computer companies.* Austin, TX.

Juliussen, K. P., & Juliussen, E. (1996). *The 7th annual computer industry almanac.* Austin, TX: The Reference Press.

Kalakota, R., & Whinston, A. (1996), *Frontiers of electronic commerce.* Reading, MA: Addison-Wesley.

Kolstad, R. (1998, January). Becoming an ISP. Available www.bsdi.com.

Kridel, D., Rappaport, P., & Taylor, L. (1997). *The demand for access to online services and the Internet* [mimeo]. Jenkintown, PA: PNR Associates.

Leida, B. (1997). *A cost model of Internet service providers: Implications for Internet telephony and yield management* [mimeo]. Cambridge, MA: MIT, Departments of Electrical Engineering and Computer Science and the Technology and Policy Program.

Maloff Group International, Inc. (1997, October). *1996–1997 Internet access providers marketplace analysis.* Dexter, MI.

Meeker, M., & Depuy, C. (1996). *The Internet report.* New York: HarperCollins.

Sidak, G., & Spulber, D. (1998). Cyberjam: The law and economics of Internet congestion of the telephone network. *Harvard Journal of Law and Public Policy, 21*(2), 327–394.

Stapleton, P. (1997). Are dial-up subscribers worth $280 per head? *Boardwatch, 11*(5), Available http://boardwatch.internet.com/mag/97/may/bwm58.html.

U.S. Department of Commerce, Bureau of the Census (1992). *Contiguous County File, 1991: US.* Ann Arbor, MI: Interuniversity Consortium for Political and Social Research.

13

Proxy Models and the Funding of Universal Service

David Gabel
Queens College

Scott K. Kennedy
Independent Analyst, Florence, MA

A few years ago U.S. state and federal regulatory commissions and local and interexchange carriers recognized the need to modify the method used to provide support to high-cost areas. Many parties agreed that there was a need to construct an economic cost model that would capture the essential factors that drive cost variations across geographical areas.

Two fundamental steps are involved in constructing an economic cost model. The model algorithms, or platform, are formulas that represent the network and network engineering. The inputs to the model help define the cost of different facilities and expenses.

Often the selection of model inputs has been based on the opinion of subject-matter experts. The opinion of subject-matter experts is difficult to validate and therefore many regulatory commissions have voiced a preference for relying on data that are in the public domain. In this chapter we use data that are in the public domain to estimate the cost of the installing cables and poles, two of the most important inputs to an economic cost model. Through econometric analysis of the data, we are able to identify how the cost of installing plant varies as a function of density and rock, water, and soil conditions. Our results have been used by state regulatory commissions to evaluate the economic cost of service, and the Federal Communications Commission (FCC) is currently pursuing a similar analysis.

THE NEED FOR NEW DATA

State regulators face many challenges and opportunities as a result of the Telecommunications Act of 1996. The National Regulatory Research Institute (NRRI) monograph on which this chapter is based was intended to provide information and data to assist regulators in meeting the challenge and opportunity created by their need to determine the cost of providing universal telephone service and the cost of unbundled network elements provided to entrants. This project began as an outgrowth of the FCC's January 1997 workshops on cost proxy models. During those meetings it was apparent that, whatever the chosen model platform, the costs used as inputs in the models should be able to be validated by regulatory commissions. This validation exercise is extremely difficult to perform on the outside plant inputs used in the Benchmark Cost Proxy Model (BCPM) and the Hatfield Model (HM) due to the fact that the model sponsors rely on proprietary data and the input of subject-matter experts whose deliberations are also not publicly available. The magnitude of these difficulties is amply illustrated by a brief analysis of how the sponsors of the BCPM model derived their outside plant estimates.

The BCPM model uses three steps to obtain its estimates of the cost of installing cables. In the first step, the model sponsors identified the cost of different activities. These data were derived from proprietary contracts between the local exchange carriers (LECs) and construction companies. The reasonableness of some of these costs was addressed in Gabel (1996).[1]

The next step indicates the model sponsors' estimate of the likelihood that these procedures will be used. In this step, the percentage estimates were established by asking outside plant engineers what they believed to be the likely mix of different activities.

In the final step, the unit cost is multiplied by the likelihood of the activity. The sum of these products is the weighted cost of installing a cable for a given density band.

The sponsors of the BCPM obtained their estimated mix of activities by discussing the matter with a team of experts.[2] To validate the claimed mix of activities, a commission or other party would need to consult with engineers that are familiar with the installation practices of the utility.

Obtaining this type of expert opinion would be difficult. An interested party or regulatory commission will find it difficult to find its own panel of outside plant engineers that could indicate the mix of activities for the different companies it

---

1. See David Gabel, *Improving Proxy Cost Models for Use in Funding Universal Service* (Columbus, OH: NRRI), 1996.

2. Contractors are paid a higher price per foot when trenching takes place in difficult situations. Identification of a difficult installation is often done either by a field inspection or by the contractor claiming that because of rocky conditions, a rock saw or blasting had to be used rather than a backhoe or plowing. The extent to which these field conditions correlate with the mix of activities identified by the models are estimated in this chapter.

regulates. For each of the regulated companies, the panel would need to know how the mix of activities varies across density zones and by soil type.

Similarly, the sponsors of the HM consulted their own group of outside plant engineers. Version 2.2.2 of the HM contained cable installation costs that were largely based on the opinion of their consultants. These consultants claimed that the mix of activities identified by the BCPM resulted in an overstatement of installing cables.

Apparently, in reaction to the criticism that the opinion of the Hatfield engineers was difficult to validate, the sponsors of the HM solicited cable cost bids from different vendors. They asked the vendors what they would charge for placing cables in different soil types. The vendors' responses exhibited significant variation, the cause of which was not clear. Some variation may be attributable to the lack of a clear definition of what constitutes a suburban or rural area, or normal versus sandy and rocky soil.[3]

Clearly the current methods for identifying the cost of installing outside plant facilities involve too many value judgments on the part of various experts. Such judgments cannot be audited and are costly, if not impossible, for independent parties and commissions to verify. It is our view that the development of cost data for use in proxy models must be transparent if the public interest is to be adequately served. With this end in mind we sought out publicly available data that could be used to obtain estimates of the cost of installing equipment.

The NRRI report, from which this chapter is drawn, uses data that are in the public domain to provide independent estimates of the cost of placing outside plant facilities and digital switching equipment. The outside plant data, which are the focus of this chapter,[4] were provided by the U.S. government's Rural Utilities Service (RUS). The RUS provides loans to independent companies to finance the reconstruction of existing exchanges and extension into new territory. A loan recipient must abide by the engineering standards established by the RUS and must issue a request for proposals if the installation is done by a contractor. The contract is awarded to the firm that submits the lowest bid and abides by the federal engineering standards.

Our data set contains information from RUS companies that operate under various soil and weather conditions, as well as population densities.[5] RUS provided recent contracts for the firms on the list and, where data were unavailable for a particular company, it provided a substitute contract. The data derived from these contracts were entered into a Microsoft Access database that is available to any in-

---

3. See "In the Matter of the Interconnection Contract Between AT&T Communications of the Mountain States, Inc. and GTE Southwest, Inc. Pursuant to 47 U.S.C. Section 252," New Mexico State Corporation Commission, Docket No. 97–35–TC, para. 47.

4. The switching estimates are available at the NRRI website identified in the acknowledgments.

terested party from the NRRI's website. This database contains 12,679 records of the unit costs of labor and materials associated with installing various outside plant facilities. Records are drawn from 171 contracts covering 57 companies in 27 states. The regression analysis was performed using the STATA statistical program.

In the interests of brevity, we focus our discussion in this chapter on the results pertaining to buried cable and pole costs. The estimates discussed here, as well as those mentioned in the NRRI report, may be used in a variety of contexts, including developing cost inputs for use in estimates of the cost of universal service and establishing the cost of unbundled network elements. They have, in fact, been used in several state dockets on unbundled network elements (UNEs) and universal service funds (USFs) as of this writing as a means of independently checking the reasonableness of outside plant figures that have been proposed by the various parties in those dockets.[6]

What this database does is to provide a set of actual construction expenditures that may be used by analysts to estimate how cost varies by the factors just identified as a means of verifying the reasonableness of factors arrived at by a panel of experts. In addition, analysts may use the databases and other supporting files located on the NRRI website to develop their own estimates of these costs.

This data set will not end the controversy about the appropriate value of inputs. This report contains a limited number of econometric specifications of the cost function. Others will, no doubt, propose different specifications that will, arguably, have superior statistical properties, make better economic and engineering sense, or result in higher or lower cost estimates.[7] It is accepted a priori that these regressions can be improved on. Nevertheless, we believe that these estimates are reasonable on the following grounds. First, based on familiarity with cost data, many of the parameter estimates appear to be reasonable. Second, almost all the models have F values that are highly statistically significant.[8] The high statistical

---

5. In performing our data collection we made every effort to identify and gather data from RUS companies operating under various soil and weather condition, as well as population densities. That said, it should be noted at the outset that RUS companies primarily serve low-density, rural areas, although some RUS companies also provide service to suburban areas. Therefore, the variability in the population densities of the service areas covered by the contracts we analyzed was not as great as we might have hoped.

6. Washington Utilities and Transportation Commission, *In the Matter of Determining Costs for Universal Service,* Tenth Supplemental Order, UT–980311(a). November 20, 1998, and Public Utilities Commission of Nevada, *In re Petition by Regulatory Operations Staff for Investigation into Procedures and Methodologies to Develop Costs for Bundled and Unbundled Telephone Services and Service Elements in Nevada,* Docket No. 96–9035, March 5, 1998.

7. For example, an analyst could add a variable that controls for variation in regional labor rates.

significance indicates that the models are doing a good job of explaining the variation in the dependent variables.

Does This Data Set Reflect Total Element Long Run Incremental Cost Standards?

The FCC and many states have determined that costing models should estimate the cost of constructing facilities to satisfy the total, rather than the incremental demand for service.[9] With regards to the loop, this involves estimating the cost of serving all customers, rather than just the additional expenditures needed to satisfy the demand of new subscribers.

We believe that these data largely reflect the cost of installing facilities to satisfy total rather than incremental demand. This conclusion is based on examination of the conditions under which an RUS company typically obtains financial assistance from the Department of Agriculture. The funding typically occurs when an RUS company concludes that it needs to rebuild its plant. The company decides that the existing facilities are no longer sufficient, and, therefore, there is a need to rebuild a portion of, or its entire, service area. Typically, RUS funding is not provided to satisfy some incremental demand, such as a new housing development. Rather the more typical case involves a rebuild of a portion or the entire service territory. During the rebuild, all pre-existing facilities are not displaced. Nevertheless, the projects are larger than required to satisfy some small increase in demand. Thus, this data set can be used either to provide inputs to a proxy model or to evaluate the reasonableness of the inputs proposed by other parties.

REGRESSION RESULTS

Model Specification

In this section, estimates of the cost of installing poles and buried cables are presented. The regression estimates were obtained through a three-step process. First the cost data from the RUS were collected. In the second stage, the cost data were

---

8. The F test is a measure of the overall significance of the estimated regression. Conducting a statistical test of the overall explanatory power of a model is much more meaningful than a claim that R^2, the coefficient of determination, is "high" or "low." What constitutes a high or low value for R^2 is subjective. This subjectivity is largely removed when statistical tests are conducted.

9. See FCC Rules, Part 51, §51.505(b), established in FCC 96–325, August 8, 1996, referring to total element long run incremental cost (TELRIC). Total means the entire production of the element, not just the next piece. A TELRIC study is different from an incremental cost study in that the cost of an element, rather than a service, is the focus of analysis.

processed and combined with information provided by the proxy model sponsors. The processing involved loading miscellaneous activities, such as splicing and street restoration, on to the cost of the cables. Also, all cost data were converted to 1997 dollars. For each of the companies in the database, data concerning line counts, area, and geological items were extracted from the BCPM 3.0, HM 4.0, and HM 5.0.[10] Finally, regression analysis was used to identify factors that explain the variation in unit costs.

The econometric models found herein varied depending on the type of facility being analyzed. The basic specification of the cost relationship was as follows:

investment per 1,000 feet = β_1 * number of pairs +β_2 * rock hardness + β_3 *soil type + β_4 * water depth + β_5 * line density + β_6 * shared installation + ε

Where:

β_1 = The incremental investment for each 1,000 pair feet of cable.

β_2 = The incremental placement investment when rock raises the cost of installation.

β_3 = The incremental placement investment when soil type raises the cost of installation.

β_4 = The incremental placement investment when water raises the cost of installation.

β_5 = The fixed cost for each 1,000 feet for placing the cable in a given line density zone.

β_6 = The reduction or increase in cost for placing one or more cables at the same location.

ε = Random error.

---

10. Certain information is taken from the HAI Model, copyright (c) 1998 HAI Consulting, Inc., AT&T Corporation, and MCI Telecommunications Corporation, and/or the BCM-Plus Model copyright (c) 1996, MCI Telecommunications Corporation. Used by permission. Certain information is taken from the Benchmark Cost Proxy Model, copyright (c) 1998 BellSouth, Indetec, Sprint and US WEST. Used by permission. Note the Hatfield Model was recently renamed as the HAI Model.

With the exception of shared installation, all these variables have a direct linkage to the proxy models. The proxy models use soil type, rock hardness, density, water depth, and the size of cable to determine the level of investment. The last explanatory variable, shared installation, is not explicitly included in the models. Rather, the economies from placing two or more cables along the same route are reflected in the estimation of the structure costs. For example, once a trench is dug for the first cable, the proxy models assume that the hole can be used by additional cables. With the parameter estimates, the cost of placing the cable should be reduced by the variable *colocate* wherever there is a second cable sharing the same trench.

The decision was made to report only the regression results utilizing the Hatfield terrain and line density data to limit the length of the study. Use of the HM should not be seen as an endorsement of that model relative to the other two proxy models, the BCPM or the FCC Staff Model, Hybrid Proxy Cost Model (HPCM).

Identical sets of regressions were run using inputs from the BCPM and the HM. The interested reader can obtain the estimates associated with the BCPM inputs by reading the STATA output file "bcpm regression log cables & poles" available on the NRRI website.

Illustration

Regression analysis was used to estimate the cost of installing different types of facilities. This analysis allows the analyst either to identify the impact of individual cost drivers or predict the cost of installing facilities when all cost drivers are considered simultaneously. This report was more interested in predicting the total cost of installing facilities rather than determining the influence of individual cost drivers. Stated differently, the primary objective is to have a specification of the cost function that provides a statistically significant explanation in the variation of the labor and material costs associated with installing facilities. Although it would be useful to have statistically significant parameter estimates for individual cost drivers, this was not the primary goal.[11]

To illustrate the regression process, the graph in Figure 13.1 identifies the cost of installing two-pair buried drop wires. The circles represent the costs reported in the contracts for 1,000 feet of buried drop wire. The X-axis variable, *combine*, identifies the soil and rock conditions.[12] A zero value for this variable implies that neither difficult soil nor rock conditions exist that increase the cost of placing the drop wire. The line running through the graph is the regression line estimated in

---

11. If individual *t* tests are conducted on each parameter estimate, caution should be taken in the construction of the confidence intervals. The hypothesis testing must reflect that if multiple hypotheses are tested, the standard Student *t* level of confidence intervals are no longer applicable. Rather, the analyst should consider using the Bonferroni test.

12. The variable *combine* is defined in the Appendix.

Table 13.1. The regression result was obtained by regressing the cost of two-pair buried cable on a constant term, _cons, and *combine*, a variable that represents soil and rock conditions.

FIGURE 13.1
Cost of Installing Buried Two-pair Drops.

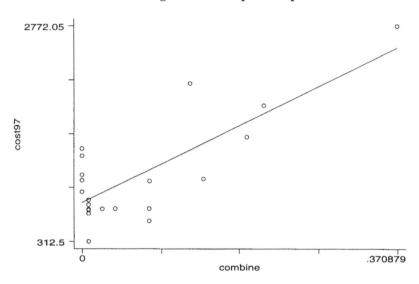

TABLE 13.1
Regression Results, Cost of Installing Two-pair Buried Drops.

```
Source |    SS    df    MS                     Number of obs = 26
--------+------------------------------      F( 1,  24) = 37.33
  Model | 4578329.18  1 4578329.18             Prob > F   = 0.0000
Residual | 2943842.18  24 122660.091           R-squared  = 0.6086
--------+------------------------------      Adj R-squared = 0.5923
  Total | 7522171.36  25 300886.854            Root MSE   = 350.23
-----------------------------------------------------------------------
 cost97 |   Coef. Std. Err.   t  P>|t|   [95% Conf. Interval]
--------+--------------------------------------------------------------
combine | 4768.784 780.5591   6.109 0.000   3157.789  6379.779
  _cons |  755.3235 80.87741   9.339 0.000   588.4007  922.2463
-----------------------------------------------------------------------
```

The sample regression function presented did not control for other factors, such as water depth and line density per square mile, that may affect the cost of installation. If these other variables are added to the specification of the model, the explanatory power of the model may increase. In the results reported in the following sections, other explanatory variables are included. This example illustrates how regression analysis can be used to identify variations in the cost of installing facilities due to differences in soil and rock conditions. These are two important cost drivers that were omitted from loop models prior to the development of cost proxy models. Prior to the introduction of these variables, models were not designed to explain why a 1,000 drop wire installed in normal soil was less costly than a similar facility installed in rocky terrain.

Buried Copper Cable

Table 13.2 provides the results from a regression analysis of buried copper cable. This represents the results from only one model estimate. Parameter estimates associated with other model specifications can be found in the file "HM regression log cables & poles," located on the NRRI website.

TABLE 13.2
Regression Results, Cost of Installing Buried Copper Cable.

```
 Source |    SS    df    MS                      Number of obs =   1131
--------+-------------------------------------   F( 6, 1125) = 996.44
  Model | 4.6588e+10   6 7.7647e+09             Prob > F    = 0.0000
Residual| 8.7665e+09 1125 7792442.79            R-squared   = 0.8416
--------+-------------------------------------  Adj R-squared = 0.8408
  Total | 5.5355e+10 1131 48943037.9            Root MSE    = 2791.5
-----------------------------------------------------------------------------

    cost97 |   Coef.  Std. Err.    t   P>|t|   [95% Conf. Interval]
-----------+-----------------------------------------------------------------
  number_o |  11.8932 .2477075  48.013 0.000   11.40718   12.37922
  colocate | -1101.283 208.6911  -5.277 0.000   -1510.751  -691.8157
   combine |  1349.091 432.9728   3.116 0.002   499.5661  2198.616
  watlin15 |  229.6622 189.3575   1.213 0.225   -141.8713  601.1958
     dens0 |  1932.864 570.1793   3.390 0.001   814.1294  3051.598
     dens1 |  2427.494 155.1588  15.645 0.000   2123.061  2731.927
```

Note. Definitions for the variables in this and other tables in the chapter may be found in the Appendix.

Note that the F value for the model is highly significant.[13] This suggests that, at any standard level of significance, the hypothesis that all of the coefficient estimates are equal to zero can be rejected. The dependent variable in this regression was cost97; when using these parameter estimates in a proxy model, there is a need to add the cost of splicing as well as LEC engineering.

These parameter estimates can be used to estimate the cost of installing buried cables. To see how this could be done, assume the following:

Cable size: 400 pairs

Presence of a second cable: no

Bedrock indicator value in HM: 1.2

Soil type indicator value in HM: 1.1

Water indicator value in HM: 1.0

Population density: 4 lines per square mile

Using these assumed inputs and the results of the regression, the following estimate would result.

Estimated investment per 1,000 feet:

$11.8932*400 - 1,101.283*0 + 1,349.091 * (1.2 - 1 + 1.1 - 1) + 229.6622$
$* (1 - 1) + 1,932.864 = \$7,094.87.$

On a per-foot basis, the estimated investment is $7.09 per foot for the 400 pair cable.

As another example, assume that the soil, water and rock type was normal, and that a 400 pair cable was used:

Cable size: 400 pairs

---

13. The variable for the depth of the water table (watlin15) is not statistically significant at any standard level of significance. The regression results from dropping this variable are found in file "HM regression log cables, drops & poles." The log file does not contain regression results from rerunning each model after dropping statistically insignificant variables. As mentioned at the outset, the objective of this study was to estimate a regression function that provided a statistically significant explanation of the variation in the value of the dependent variable.

Presence of a second cable: no

Bedrock indicator value in HM: 1.0

Soil type indicator value in HM: 1.0

Water indicator value in HM: 1.0

Population density: 4 lines per square mile

Using these assumed inputs and the results of the regression, the following estimate would result:

Estimated investment per 1,000 feet:

$11.8932*400 - 1,101.283*0 + 1,349.091 * (1 - 1 + 1 - 1) + 229.6622 *$
$(1 - 1) + 1,932.864 = \$6,690.14.$

On a per-foot basis, the estimated investment is $6.69 per foot for the 400 pair cable. These two examples highlight how the cost of installing cables varies depending on rock, soil, and water conditions. In the second calculation, where the soil and rock conditions were normal, the sample coefficient estimates suggest that the cost of installing the cable is lower by approximately 40 cents per foot.

Buried Fiber Optic Cable

Table 13.3 provides the results from a regression analysis of buried fiber optic cable. Results are reported for only one model estimate. Parameter estimates associated with other model specifications can be found in the file "HM regression log cables, drops & poles."

In order to see how to use these parameter estimates, consider the following assumptions:

Cable size: 16 tubes (strands of fiber)

presence of a second cable: no

bedrock indicator value in Hatfield Model: 1.0

Soil type indicator value in Hatfield Model: 1.0

Water indicator value in Hatfield Model: 2.0

Population density: 25 lines per square mile

Using these assumed inputs and the results of the regression, the following estimate would result.

Estimated investment per thousand feet:

$36.58157*16 + 56.43567*0 + 593.2549 * (1.0 - 1 + 1.0 - 1) + 121.8819 * (2 - 1) + 2,316.261 = \$3,023.45.$

On a per-foot basis, the investment is $3.02 per foot for the 16-fiber cable. As with all cable regression estimates, there is a need to add the cost of LEC engineering and splicing.

TABLE 13.3
Regression Results, Cost of Installing Buried Fiber Optic Cable.

Source	SS	df	MS		Number of obs = 707
					F(6, 701) = 589.26
Model	1.0212e+10	6	1.7019e+09		Prob > F = 0.0000
Residual	2.0246e+09	701	2888212.72		R-squared = 0.8345
					Adj R-squared = 0.8331
Total	1.2236e+10	707	17307154.4		Root MSE = 1699.5

cost97	Coef.	Std. Err.	t	P>|t|	[95% Conf. Interval]
number_o	36.58157	1.909966	19.153	0.000	32.83163	40.33151
watlin15	121.8819	155.8343	0.782	0.434	-184.0759	427.8397
combine	593.2549	296.9037	1.998	0.046	10.32789	1176.182
dens0	1850.096	347.5987	5.323	0.000	1167.637	2532.555
dens1	2316.261	124.1455	18.658	0.000	2072.52	2560.003
colocate	56.43567	146.5545	0.385	0.700	-231.3027	344.1741

The incremental cost per pair foot of fiber strands, $0.03658*2$, is significantly greater than the cost of a pair of copper cables, $0.01189. On the other hand, the fixed costs per pair foot of installing buried fiber and copper cables are similar. Dens0 and dens1 measure the fixed cost per foot of placing a cable in the 0 to 5, and 6 to 100 lines per square miles density zones, respectively. The confidence interval estimates for variables dens0 and dens1, as reported in Tables 13.2 and 13.3, overlap. The similarity in placement costs, along with fiber's greater capacity, explains, in part, why fiber cable is increasingly the preferred facility in the feeder portion of the network.

Poles

The database contains information on the cost of various sizes and classes of poles. The regression analysis provided in Table 13.4 are estimates for a 40-foot, Class 4 pole. This size pole was used because this is the size pole the HM assumes. We were unable to identify from the BCPM documentation the assumed pole size, but we believe the model uses 45-foot, Class 5 poles.

TABLE 13.4
Regression Results, Investment per Pole.

```
    Source |    SS      df    MS                   Number of obs =    19
-----------+------------------------------         F( 3,   15) =   3.51
     Model | 61704.4186   3 20568.1395             Prob > F     = 0.0415
  Residual | 87822.3628  15 5854.82419             R-squared    = 0.4127
-----------+------------------------------         Adj R-squared = 0.2952
     Total | 149526.781  18 8307.04341             Root MSE     = 76.517

-------------------------------------------------------------------------------
    cost97 |    Coef.   Std. Err.    t    P>|t|    [95% Conf. Interval]
-----------+-------------------------------------------------------------------
  sstare15 | 49.99036  244.7599   0.204  0.841   -471.703    571.6837
  watlin15 | 112.5506   50.40416  2.233  0.041    5.116637   219.9845
  bed48a15 | 66.07799   31.45088  2.101  0.053   -.9579629   133.1139
     _cons | 310.645    33.64669  9.233  0.000    238.9288   382.3612
```

All of the 40-foot, Class 4 observations appear in the Hatfield density band of 5 to 100 lines per square mile. Consequently, the regressions given do not include dummy variables for different density zones. Note that in neither the BCPM nor the HM do the model sponsors contend that the cost of a pole is a function of the population density.[14]

To see how to use these parameter estimates, consider the following assumptions:

Soil type indicator value in HM: 1.0

Water indicator value in HM: 2.0

Bedrock indicator value in HM: 1.0

Using these assumed inputs and the results of the regression, the following estimate would result:

Estimated investment per pole:

$49.99036*(1.0 - 1) + 112.5506 * (2.0 - 1) + 66.07799 * (1.0 - 1) + 310.645 = \$423.20.$

If the water depth indicator value was also one, the cost of the pole would be $310.645.

Note that the pole costs do not include the cost of guys, cross arms, grounding wires, anchors, or other miscellaneous items. Also, as with all of the RUS outside plant data, LEC engineering is not included. The RUS Access database contains many items that are associated with the cost of placing poles. This list of miscellaneous pole-related items, their cost, and the number of times they appeared in the database was faxed to several experts in the area of outside plant engineering, who were asked to provide an opinion as to how these miscellaneous items should be factored into pole costs. Based on the comments of the experts, pole loading calculations were devised.[15]

The results of our analysis suggest that the pole loadings for rural, suburban, and urban areas are $32.98, $49.96, and $60.47 per pole, respectively. It should be noted that there was some disagreement around the issue of ground wire assembly and pole grounding units. One expert felt that the probability estimates for these units were too low. It was his opinion that, especially in high-lightning areas, each pole would have a pole ground assembly unit attached to it, and that there would be commensurate increase in the number of other grounding units. Although we believe that the percentages used in the table are reasonable, analysts may change these percentages either to reflect the conditions that exist in their specific area of

14. HM 5.0 Inputs, Assumptions, and Default Values, B12—Pole Investment, December 11, 1997; and BCPM 2.0, file loopint.xls, folder Structure Inputs.

15. The calculations supporting this table may be found in the *Pole Loadings Calculation Workbook.xls* file, available on the NRRI website or from the authors.

the country or to conform with the engineering practices employed by the companies under consideration.

QUALIFICATIONS

The two major qualifications concerning our study that we discuss in this chapter are those concerning the applicability of the RUS data to other density zones and the issue of structural sharing. For a fuller discussion of other issues, such as loading for telephone company engineering, the reader is directed to consult the full NRRI report mentioned at the beginning of this chapter.

Other Density Zones

Because RUS companies generally serve rural, low-density areas, the RUS data set only contains outside plant data for the three lowest density zones. Therefore, the data provide only limited insights into the cost of serving urban territories. For USF studies this does not present a problem as the vast majority of USF support is targeted to the density zones covered by the RUS data. For UNEs a similar study will have to be done for the higher density zones. This is not to say, however, that the RUS data as they exist could not be used to estimate the cost of serving urban territories.

For example, an analyst could use the regression estimates to forecast costs that would be incurred in more densely populated areas. As a matter of sound econometrics, however, caution must be exercised when parameter estimates from a data set are used to forecast costs for areas that are too dissimilar to those from which the data were obtained.

Alternatively, an analyst could use the RUS data set to analyze a proxy model's inputs for rural and suburban areas. If, for example, the data from the RUS contracts are 15% higher than the inputs to a proxy model, the proxy model's inputs for urban areas could be adjusted upward by 15%. The underlying hypothesis of such an adjustment would be that the error for rural and suburban areas is systematic and, therefore, the same magnitude of adjustment should be made for urban areas.

Sharing

Sharing of structural investment occurs when one or more utilities share the cost of placing outside plant facilities. For example, electric and telephone companies may use the same pole to place their cables; or telephone, electric, and cable television companies may place their cables in a common trench. When sharing does occur, the cost of placing telephone plant is reduced.

The cost proxy models allow the user to declare the extent to which facilities are shared. For example, a user could declare that only 50% of the cost of installing poles is recovered from telephone operations.

This report presents data on the cost of installing outside plant facilities. The contracts do not state the degree to which the final contract prices reflect sharing. This section explains how these contract prices should be further adjusted to reflect sharing.

Buried Cables

The contracts show that, in the service territory of RUS companies, buried cable is the predominant mode of cable facility. The contracts also indicate that the vast majority of the buried cable is plowed rather than trenched.[16]

If the cost of placement is borne exclusively by the telephone company, the cost of plowing in normal soil is less than the cost of trenching. When cables are placed through plowing, the cost is rarely shared with other utilities.[17]But to the extent sharing does occur, the buried cable costs reported in this database largely reflect the costs incurred by LECs after taking sharing into account.

There is one area in which this data set does not reflect sharing. When a developer incurs the cost of opening up the ground, multiple utilities place their cables or pipes into the pit. Because the cost of opening the ground is borne by the developer, rather than the utilities, the cost of this activity would not be included in the data set. To take this type of situation into account, assignment of less than 100% of the buried structural costs to telephone operations merits consideration.

Poles

The contracts indicate the use of many different sizes and classes of poles. The higher the pole, the greater the likelihood that it is shared by multiple utilities.

The proxy models assume that a 40-foot, Class 4 pole is installed. The contracts do not indicate whether any of the pole costs have been shared with other utilities. It is our understanding that the poles are owned exclusively by the

---

16. When the cable is placed in a trench, the RUS contracts require a "T" designation. Column I, "Trenching Indicator," indicates that trenching is not widely used by the RUS contractors. To some extent, cables placed with trenching may not be properly identified either in the contracts or the database.

17. See, for example, *In the Matter of the Pricing Proceeding for Interconnection, Unbundled Elements, Transport and Termination, and Resale,* Washington Utilities and Transportation Commission, Docket No. UT–960369, Tr. Vol. 10 (July 8, 1997), pp. 323–326.

LEC. Therefore the costs recorded in the database do not reflect any sharing of pole structural investment.

CONCLUSION

Comparison of RUS With HAI and BCPM Data

Table 13.5 illustrates no general pattern between the proxy model estimates and the costs reported in the RUS contracts. Our research suggests that the proxy models sponsors have understated the cost of placing buried cables and have overstated the cost of installing poles. As already suggested, the advantage of the data contained in this study is that we have relied on contract values, rather than the opinion of subject-matter experts.

TABLE 13.5
Comparison of RUS with HAI and BCPM Suggested Inputs

	Normal soil	Soft rock	Hard rock
Pole cost comparisons			
HAI model values	417.00	834.00	1,469.50
BCPM model values	726.75	726.75	926.75
Gabel-Kennedy RUS values	341.71	414.50	487.08
Buried structure values for the 0 to 5 lines per square mile density zone			
HAI model values	1.77	2.66	3.98
BCPM model values	1.41	2.16	3.56
Gabel-Kennedy RUS values	1.69	3.17	4.60
Buried structure values for the 6 to 100 lines per square mile density zone			
HAI model values	1.77	2.66	3.98
BCPM model values	1.75	2.37	3.95
Gabel-Kennedy RUS values	2.23	3.72	5.20

How Have These Data Been Used?

An earlier version of this data set was recently provided to parties in a proceeding before the Maine Public Utilities Commission. The hearing examiners in an unbundled network element case asked the parties to comment on the outside plant costs. A Bell Atlantic (BA) witness submitted a comparison of the material costs and material and labor costs from the RUS data with the comparable cost inputs used in BA's link (loop) study. His principal conclusion was "that the material costs used by the Company in the TELRIC Study are less, for all cable types, when compared with the RUS data."[18]

In a USF docket in Washington State, the State Commission utilized the RUS outside plant numbers to verify the outside plant numbers utilized by the various parties to the proceedings in making in their proxy model runs.[19]

In Nevada, the State Commission adopted the RUS outside plant numbers and our switch cost numbers as default values for use in the HM, which the Commission had adopted for use in its USF and UNE dockets.[20] In making that decision the Commission felt that the RUS numbers provided a reasonable substitute for the HM default values and were valuable in providing the Commission with actual, publicly available outside plant and switching costs as a check on the reasonableness of the company-proposed values for those costs.

Cost proxy models are being used as a regulatory tool in universal service fund and unbundled network element dockets around the country. A significant challenge to state commissions considering these models has been the validation of the cost data that are used to populate them. As mentioned earlier, this validation can be costly and very difficult, if not impossible, for state commissions to perform on their own. The RUS data discussed in this chapter provide commissions with publicly available data, derived from actual contracts, on the cost of placing outside plant facilities. As we have shown, these data are a valuable tool that can be, and have been, used by commissions as a means for judging the reasonableness of the cost estimates submitted to them.

In fact, the FCC has initiated a study with the intent of obtaining data similar to what we have obtained through the RUS contracts. The notice for this study is entitled *Common Carrier Bureau Requests Outside Plant Structure and Cable Cost*

---

18. Rebuttal testimony of Stanley Baker, Bell Atlantic, Docket No. 97–505, December 22, 1997, p. 4, and Attachment 1. Baker also found (at pp. 4–5) "[i]n contrast with the comparison of material costs which showed that the Company's studied costs were lower than the RUS data, the comparison of labor and material costs between the RUS data and the Company's data shows just the opposite" (Ibid.). Among other factors, Baker notes that "[o]ne item causing this result is the labor component associated with fiber splices in the independent company data is inordinately low." Baker also stated (at pp. 5–6) that BA's underground copper costs were high relative to the RUS values because "The Company's costs for underground copper cable reflect only the relatively short lengths deployed in the TELRIC study." A fuller discussion of how these comparisons may be used, in conjunction with the RUS data, in estimating the forward-looking cost of a large firm is contained in the authors' NRRI report.

19. Washington Utilities and Transportation Commission, *In the Matter of Determining Costs for Universal Service,* Tenth Supplemental Order, UT–980311(a). November 20, 1998.

20. Public Utilities Commission of Nevada, *In re Petition by Regulatory Operations Staff for Investigation into Procedures and Methodologies to Develop Costs for Bundled and Unbundled Telephone Services and Service Elements in Nevada,* Docket No. 96–9035, March 5, 1998.

Data for High Cost Model (available on the FCC website: www.fcc.gov/Bureaus/ Common_Carrier/Reports/survey.wp).

The data contained in the RUS database can be easily updated using price indexes. However, we believe that gathering more publicly available data that can serve to augment this database, as the FCC is doing through its study, is the most useful way to go. This will ensure that state commissions have a variety of up-to-date cost data to assist them in the challenging task they face in determining the cost of providing the set of telephone services included in universal service and in determining the cost of unbundled network elements provided to new entrants under the Telecommunications Deregulation Act of 1996.

ACKNOWLEDGMENTS

This chapter is based on our April 1998 NRRI at the Ohio State University manuscript entitled *Estimating the Cost of Switching and Cables Based on Publicly Available Data*. NRRI kindly provided financial support for the project. Those wishing to read the report and obtain the database used in creating the report may obtain them from the NRRI website (http://www.nrri.ohio-state.edu/). The report may be found by clicking on the *New Research Available for Download* option and is on that Web page as *98-09 Estimating the Cost of Switching and Cables Based on Publicly Available Data.*

Many individuals helped us obtain information for this chapter. Our gratitude extends to many, but we want to give special thanks to Gary Allan and Ed Cameron of the RUS of the U.S. Department of Agriculture.

APPENDIX A: DEFINITIONS OF REGRESSION VARIABLES

Variable	Definition
bed36115	Hatfield 5.0 hardness indicator value at 36-inch depth – 1. CBGs weighted by lines.
bed48a15	Hatfield 5.0 rock hardness indicator value at 48-inch depth – 1. CBGs weighed by area.
colocate	Two cables simultaneously placed at the same location.
combine	Bedrock + soil indicator = bed36115 + sstlin15 Hatfield value. 36 inches is used for copper and fiber because this depth appeared most frequently in the contracts.
_cons	Constant term in STATA regressions.
cost97	Cost of facilities, expressed in 1997 dollars, including appropriate loadings, with one exception. For cables, splicing costs are excluded.
dens0	0 to 5 lines per square mile. Calculated by dividing Hatfield lines by Hatfield area.
dens1	6 to 100 lines per square mile. Calculated by dividing Hatfield lines by Hatfield area.
number_o	Number of copper pairs or fiber strands in cable sheath.
sstlin15	Hatfield 5.0 soil surface indicator value – 1. CBGs weighted by area.
sstare15	Hatfield 5.0 soil surface indicator value – 1. CBGs weighted by area.
watlin15	Hatfield 5.0 water indicator value – 1. CBGs weighted by lines.

IV

CONVERGENCE

Internet Telephony or Circuit Switched Telephony: Which Is Cheaper?

Martin B. H. Weiss
Junseok Hwang
University of Pittsburgh

Telephony is an Internet application that has the potential to radically alter the telecommunications environment. This application may affect traditional regulatory structures, subsidy structures, business models, and so on. Today, users can transmit telephonelike voice traffic over the Internet at zero incremental price, unlike circuit switched telephony, which usually has a per-minute incremental price. This research was carried out to determine whether Internet telephony (Itel) was a fundamentally cheaper approach to the interoffice transmission and switching of voice, or whether the price difference was the result of an implicit regulatory subsidy.

This chapter is intended to be a first estimate of the switching and interoffice transmission costs for Itel and circuit switched networks. To evaluate these costs, we compare "greenfields" networks in both technologies, dimensioned for an area and a population equivalent to the state of Rhode Island. We find that the switching and transmission costs for Itel are approximately 50% lower than the costs for circuit switching. We further find that this cost difference is largely due to the reduced interoffice transmission capacity required by Itel and not asymmetrical regulatory treatment.

INTRODUCTION

One of the most challenging developments in telecommunications in recent years has been the emergence of Itel. In the United States, Internet service providers (ISPs) are exempt from the access charge system that is used to support local service. Internationally, ISPs are often outside of the traditional regulatory structures because they are value-added networks (VANs), which have historically been less regulated than providers of public switched service. Furthermore, the international accounting and settlements process is substantially challenged in the face of Itel, because the benefits of arbitrage are substantial. Even if it were a social goal to include Itel under the normal regulatory framework, how to do so is far from clear (Frieden,1997).

The question is frequently asked: Is Itel's per-call price lower than the public switched telephone network's because Itel is fundamentally more cost effective, or because of a regulatory artifact?[1] If it is the former, it implies a pending revolution in the design, organization, and operation of the public network infrastructure. If it is the latter, Itel will be a marginal phenomenon in the long run, requiring little if any attention from public regulators.[2]

In this chapter, we begin to address this question by constructing networks (on paper) and evaluating their costs. This is not intended to be a complete or final design, but rather one that captures the major switching and transmission cost elements of the interoffice network needed to address the question articulated in the previous paragraph. This is intended to be a first-order analysis; there are many assumptions, the relaxation of which may result in further insights.

TECHNOLOGICAL OVERVIEW

The purpose of this section is to outline the key distinctions between these two technologies. More detailed technical discussions can be found elsewhere.[3]

Assumptions and Simplifications

Given that this is a first-order analysis, we make the following simplifying assumptions:
- *The focus is on the costs of switching, signaling, and trunking.* Thus, we assume that similar access and transmission technologies will be used. An actual Itel-based network might well consider alternatives to the current local loop technology. The assumption about transmission is not unreasonable, since the higher speed aggregate transmission links, such as SONET/ SDH and DS3, would use the same technology.

---

1. This sentiment has recently been expressed by Jack Grubman, a well-known telecom analyst with Salomon Smith Barney (in Stuck & Weinggarten, 1998).

2. Investments by carriers such as Qwest Communications suggest that the cost may be lower.

3. See Kostas et al. (1998), Kumar, Lakshiman, and Stiliadis (1998), Low (1997), Manchester, Anderson, Doshi, and Dravida (1998), and Schulzrinne (1997).

- *We assume current technologies.* We used compressed voice and silence suppression technologies for the Itel access networks and fast Internet protocol (IP) Layer 3 switches (sometimes called gigabit switch routers) for Itel switching functions.
- *We assume equal levels of demand for both technologies.*
- *We assume that the services are perfect substitutes for each other.* That is, we assume that the user will not be able to tell the difference between Itel and traditional telephone service from the point of view of major functions. Today, many consumers report poorer service quality with Itel as well as limitations surrounding the computer (Clark, 1998). On the other hand, Itel allows a level of service integration that is difficult or costly to achieve with circuit switched telephony.
- *We assume that neither service is subject to line charges for regulatory purposes.*
- *We assume this network is only connected to similar networks.* As a result, we make no allowance for gateway or interconnection facilities.
- *We assume that the cost of transmission is constant over the life of the study.*

The Public Switched Telephone Network

The public switched telephone network has evolved into its present form over its 100-year history (see Fig. 14.1). The network was initially optimized to handle low-bandwidth (4kHz) channels using manual technology (no mean feat, as illustrated by Mueller, 1997). As technology evolved, so did the way in which switching was performed. The digitization of the network allowed for high-speed data services. Advances in packet switching technology allowed for the transformation of the signaling network to support a wide array of enhancements to basic service (these are implemented via the signal transfer points [STPs], signal switching points [SSPs], and service control points [SCPs] in Fig. 14.1). Despite these advances, the circuit switched telephone network can be characterized by the following:

- It is capable of handling many dedicated low-bandwidth channels (64kbps or 4kHz). Adaptive Differential PCM (ADPCM) was developed to transmit "toll quality" voice over 32kbps, but this technology has not been widely installed.
- It is independent of content: Once the channel has been allocated, it remains allocated whether it is used or not; that bandwidth cannot be used by others during idle periods.
- Network attachments (i.e., telephones) are cheap because their functionality is specialized and limited.

Numerous other characteristics also exist; this list attempts to capture those of relevance to this study.

FIGURE 14.1
Circuit Switched Architecture.

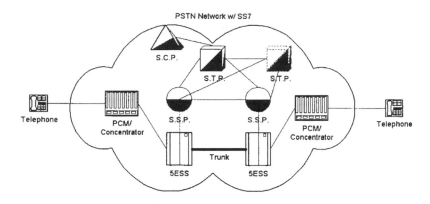

Itel

Itel has grown as a specialized application of the Internet (see Fig. 14.2 for an illustration). A dominant characteristic of TCP/IP, as with most packet networks, is that most resources are shared (as opposed to dedicated). Thus, the bandwidth of a transmission channel is dynamically allocated to those who are using it at the moment. If their use disappears for a time, no system resources are dedicated to that user. The system was not designed to support services that require guaranteed timely packet arrivals. In summary, the essential characteristics of TCP/IP networks are:

- It is capable of handling many application types and allocating bandwidth dynamically among them on demand. This makes the development of integrated services particularly easy.
- It cannot easily make performance guarantees, especially arrival time guarantees. This can lead to quality of service degradation if the network becomes congested while being used for voice traffic. Note that this problem can be substantially mitigated if the network is engineered to low utilization, but at the expense of increased cost.

Architecture of Itel

In this section, we discuss the components of the Itel architecture that differentiate it from the circuit switched network architecture. These are:

FIGURE 14.2
Internet Telephony Architecture.

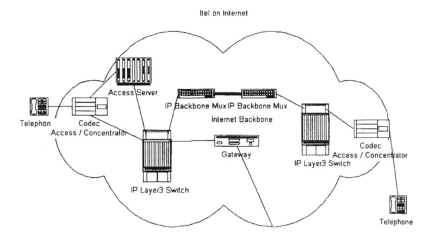

LOOP INTERFACE

Since we assumed only that the copper wires in the loop were constant, we configured the Itel approach with xDSL. As a result, the Itel configuration includes digital subscriber line access multiplexers (DSLAMs), which raise the capital cost of Itel significantly. We further must assume that users have a means to access the Itel network; this may be a specialized device, such as an Internet telephone,[4] or software and systems running on the personal computer that implement the G.729a compression and RTP/UDP/IP protocol standards. We do not add the cost of these devices to our analysis.

IP LAYER 3 SWITCHES

The latest IP routers operate at much higher performance than traditional IP routers. They combine Layer 2 switching and Layer 3 routing in such a way that they can still interoperate with conventional IP routers but process and forward packets much more quickly.

Many recent IP switches provide packet processing rates of more than 1 million packets per second (MPPS). In our simulation, we assumed Cisco equipment (see Kumar et al., 1998, for more detail).

---

4. See, for example, http://www.selsius.com: Selsius phone 12S Series' price ranges from $200 to $400 at the time of this writing.

BACKBONE OC-3 TRUNK CARRYING IP

There are many ways of carrying IP traffic over the synchronous optical network (SONET). Some approaches place IP traffic directly onto SONET links, which provides for lower overhead and, therefore, more efficient transmission. Other approaches place IP packets in asynchronous transfer mode (ATM) cells, which are then carried over SONET. This latter approach allows the IP network to take advantage of ATM services, albeit at the cost of lower efficiency. Figure 14.6 shows the protocol stack of Itel central switching office. In our model, we place the IP packets directly on the OC-3 SONET trunks. Our simulation shows that these trunks are normally utilized less than 40% of capacity.

BACKBONE ITEL GATEWAY SUPPORTING SS7

We include a SS7 interface in our model so that this network can communicate with other networks using existing standard protocols.

CALL PROCESSOR

Routers and switches are designed to carry traffic; they do not provide the resources necessary to set up and tear down calls, provide call services (such as billing and calling features), and so on. Generally, public communications networks have a separate call processor for this task. We included a single processor Sun UltraSPARC to process calls. This may well be too small to handle the required processing loads. Because carrier-level call processing software for Itel is not yet available, we were unable to size this correctly. Information on the processing capability of 5ESS or DMS100 processors was not available to us. Thus, the cost of this element may end up being higher, but it is small compared to the cost of switching and trunking.

RTP/UDP/IP Protocol Stack in Itel

The dominant standard for transmitting real-time data in packet switched networks is International Telecommunication Union (ITU) standard H.323, which uses RTP/UDP/IP encapsulation. These protocols provide the necessary information so that Itel packets can reach their destination. They add 40 bytes of overhead to every packet; the voice information can consist of 10 or 20 bytes of information, depending on the implementation (we assumed two 10-byte payloads for this chapter).[5] Real-time transport protocol (RTP) supports end-to-end delivery services of applications transmitting real-time data over IP networks (see Fig. 14.3). RTP is defined in RFC 1889 (Casner, Frederick, Jacobsen, & Schulzrinne, 1996) and applied to audio and video in RFC 1890 (Schulzrinne, 1994). RTP does not guarantee timely delivery or quality of service. RTP typically runs over the User Datagram Protocol (UDP) to utilize its multiplexing and check-sum services. RTP provides the sequence number and time-stamp information needed to assemble a real-time data stream from packets. RTP also specifies the payload type to assign multiple data and compression types.

FIGURE 14.3
RTP Protocol Stack.

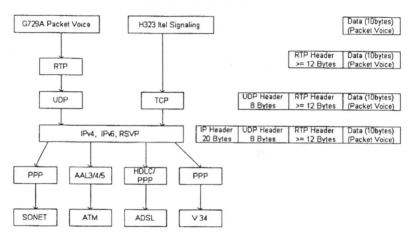

MODEL DESCRIPTION

In this section, we describe the parameters of the simulation model. The network parameters for the circuit switched case and the Itel case are presented in their respective sections along with the summary results from the simulation.

The Service Area

To estimate the cost for each system, we constructed a "greenfield" system of each type for the same service area (a population of 1 million people uniformly distributed over an area of 3,140 square km—equivalent to the state of Rhode Island; see Fig. 14.4). To simplify the calculation, we make the following assumptions:
   • An average of 2.2 people per household.

---

5. G.723 , G.729, and G.729A are the popular compression types for the codecs in Voice Over IP (VOIP). The VOIP standards committee proposed a subset of H.323 for audio over IP. Many Itel vendors developed the products based on this standard.

- Square service area (56 km per side).
- No geographical barriers.
- Households uniformly distributed over the service area (constant population density).
- Homogeneity of users.
- Consistency of a user's behavior between systems (i.e., we assume away price and demand issues and assume the same demand patterns for plain old telephone service and Itel).

FIGURE 14.4
Simulation Topology.

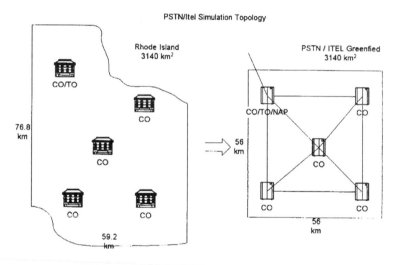

We assumed the local loop with 19 gauge copper twisted pairs that can be extended up to 30 k-ft.[6] We use five local switching locations because a 5ESS-2000 can support up to 100,000 loop lines. If we assume one line per household, then each local switching location (CO) terminates 90,909 lines. We further assume that all originating traffic is distributed as 10% for outgoing from its service region and the remainder evenly among the five COs. The incoming traffic from the other service regions is assumed to be same as the amount of outgoing traffic, passing through the tandem (or NAP).

---

6. We realize that nobody actually uses 19 AWG cable. However, this allowed us to assume "home runs," avoiding the complications introduced by loop carrier systems. Because we are not costing this portion of the network, this assumption is inconsequential.

Circuit Switched Model

In addition to the general assumptions already described, we have made the following additional assumptions (summarized in Table 14.1) that pertain specifically to the circuit switched network:

- A 5ESS-2000 switch supports 100,000 lines at 3.6 CCS/line (0.1 Erlang) in the busy hour. Therefore, each switch carries 9090.9 Erlangs of traffic from its local loops.
- We assumed that most of the blocking occurs not in the switches but at the trunk side at 1% blocking probability.
- Because we have only five switches within the given service region, we assumed each switch will be connected to every other switch via a SONET link.
- The signaling network (SS7 Signaling System 7) is configured so that each local switch is equipped with SSP and connects to two STPs and an additional two STPs, forming a quadruple mesh STP network to access SCPs and external STPs (illustrated in Fig. 14.5). Each signaling link (between SSP and STP) is engineered to have 40% utilization, so that if a failure occurred, the expected utilization would be 80% per link (Bellcore, 1987). The data rate for each signaling link is 64 kbps. We assume that each call generates on average 3.5 signaling messages from an originating party and 3.5 signaling messages from the terminating party. The average packet length is 15 octets per message.

Based on these assumptions we dimensioned the circuit switched network as follows:

- The capacity of the trunks between any two normal COs and between a CO and Tandem will be bidirectional 210,816 Kbps (for each direction) and 326,400 Kbps, respectively. With this configuration, the COMNET III simulation produced a P.01 grade of service.
- The capacity of the signaling links between a SSP and a local STP; between two local STPs; between a local STP and remote STP and between two remote STPs are bidirectional 326 Kbps (326 Kbps for each direction), 1,630 Kbps, 111 Kbps, and 222 Kbps, respectively.

The summary of circuit switched network simulation output is in Table 14.2. The simulated trunk capacity (211 Mbps) between each CO and toll office requires two OC-3 and one OC-1 SONET trunks.[7] The trunk capacity (327 Mbps) between each CO requires one OC-3 and one OC-1 SONET trunk.[8] The OC-1s shown in the table are the separate trunks between each CO, which cannot be aggregated to OC-3s.

---

7. A total of four pairs of this type are required in the simulated service area networks.

8. A total of six pairs of these types of trunks are required in the simulated service area networks.

TABLE 14.1
Assumptions for Circuit Switched Model.

Circuit switched and SS7 network parameters	Values
Circuit switched parameters	
Data rate per channel	64 Kbps
Local loop	19 gauge twisted pair
Circuit switch	5ESS-2000
Fraction of outgoing call	0.1
Originated traffic per line	0.1 Erlangs
No. of loop lines per CO	90,909
Interarrival time for CO	0.0264 sec
Interarrival time for toll office	0.0176 sec
Average call duration	667.48
SS7 Parameters	
No. of SSP	5
No. of LSTP	2
No. of RSTP	2
No. of SCP	1
Engineering utilization	40%
Unit of SS7 link group	8 DS0s
Average signaling messages per call	7
Average signaling message size	15 octets

Itel Model

The Itel simulation model assumed in our analysis is based on RTP/UDP/IP standardized in ITU H.323. In this protocol, sequence numbers and time stamps are used to reassemble the real-time voice traffic, although this provides no quality guarantee. TCP/IP is used to control the call (like SS7 in the circuit switched model). The simulated Itel model would provide functionality comparable to the circuit switched network, but it will not necessarily provide quality and reliability comparable to circuit switching.

Figure 14.6 represents the Itel architecture model in our simulation. The simulation uses IP router switching and trunking. The IP access server and edge concentrator in the figure represent DSLAMs that do not have much effect on delay in the simulation.

The assumptions made for the Itel simulation model are specified here and in Table 14.3.

TABLE 14.2
Summary of Simulation Output for Circuit Switched Network.

Circuit switched and SS7 network requirement	Values
Circuit switched parameters	
Four trunks between each CO and a toll	4 X (2 OC-3 and 1 OC-1)
Six trunks between each CO	6 X (1 OC-3 and 1 OC-1)
Blocking probability	0.01
Calls attempted per hour	750,299
Calls carried per hour	742,454
SS7 Parameters	
SS7 links between SSP and LSTP	80 DS0
SS7 links between LSTPs	32 DS0
SS7 links between LSTPs and RSTP	8 DS0
SS7 links between RSTPs	8 DS0

- All voice traffic would be compressed from 64 Kbps PCM voice to 8 Kbps compressed data using G.729A codec.
- Silence suppression will be enabled in each codec, with 60% of a session being silent in one way.
- On the suppressed codec output, RTP, UDP and IP overhead will make actual average throughput around 14 Kbps.
- Each voice is packetized every 10 msec making 10 bytes voice packet through compression codec, and two voice packets (20 bytes payload) are enveloped in the 40 bytes RTP/UDP/IP header. The operational details of the RTP/UDP/IP are described in the previous Itel section.
- For the packet voice, the burst packet voice is modeled with average 350 msec exponentially distributed active state and 650 msec exponentially distributed silence state (see Fig. 14.7).
- The Itel call is modeled as a connectionless UDP/IP session with exponentially distributed session lengths with a mean of 240 sec.

In the simulation, we found the average and 95 percentile delays for packets through the network. This delay was modeled using the delay budgets in Table 14.4. We assumed that this would provide a connection equality that was approximately equivalent to circuit switched voice[9] (Table 14.5 shows the ITU recommendations). The source was initially modeled using an "on-off" speaker

---

9. Total delay is a function of operating system and sound device delay in the end devices as well as network delay. Informal measurements suggest that these delays can be very high in WinTel computers using standard sound cards. These are not included in our delay simulations.

model. We found that, when aggregated, these could be modeled reasonably well by an exponential distribution, so that is what we used to reduce the running time of the simulation.

TABLE 14.3
Assumptions for Itel Simulation Model.

Itel networks parameters	Values
Compressed peak data rate per channel	8 Kbps (G.729A)
Local loop	19 gauge twisted pair, ADSL
Packet switch	IP switch (>10 Gbps and 1 MPPS)
Fraction of outgoing call	0.1
Originated traffic per line	0.1 Erlangs
No. of loop lines per CO	90,909
Packet voice size	10 bytes (10 msec)
Packet payload size	20 bytes
Protocol overhead	40 bytes (RTP/UDP/IP)
Packet voice burst distribution	Burst 350 msec, silence 650 msec
Packet delay constraint	Less than 250 msec
Itel call (RTP session) setup delay constraint	Less than 1 sec

Unlike a conventional ISP, which terminates the user line through a channel service unit (CSU) and data service unit (DSU) on a local area network (10 Mbps or 100 Mbps LAN), the Itel user access line for the carrier solution is terminated through a DSLAM, which is connected to the IP switches via internal OC3 or T3 circuits. Figure 14.6 shows the operational architecture of the Itel CO model. The simulation shows that 12 DS3 interfaces are required for the local interface with the concentrator from the IP switches and OC-3 line can support all trunk connection between any COs with more than 50% spare capacity. With OC-1, the trunks are so fully utilized that it increases the packet delay significantly in the simulation. Therefore, one SONET OC-3 ring provides enough capacity for the assumed Itel trunk traffic among the ISP COs. In this simulation model, SS7 call setup functions are simulated as session setup using TCP/IP. Therefore, no additional capacity dimensioning is required for each links. The output of simulation for Itel is summarized in Table 14.4.

ESTIMATED SYSTEM COSTS

With these designs in hand, we consulted with vendors to review the "reasonableness" of the design and to estimate the cost of the switches. The cost estimates we used were based on information extracted from the literature (circuit switched network) and from Cisco (Itel network). The cost of the transmission links is based on leased line costs from AT&T.

TABLE 14.4
Summary of Simulation Output for Itel.

Itel network requirement	Values
Itel networks	
Trunk among COs	OC-3 Sonet Ring (=10 OC-3)
Local interface of each IP switch	12 DC-3 ATM
Major network components	Itel access server Concentrator IP switch Sonet Mux
Itel delay budget	
Average switch utilization	0.62
Average variable packet delay	3.41 msec
95 percentile variable packet delay	18.44 msec
Access, look-ahead, encoding, dejitter G729A	77 msec
Average end-to-end packet delay	80.41 msec
95 percentile end-to-end packet delay	95.44 msec
Average delay jitter	8.45 msec

The cost shown here is only focused on switching and trunking for circuit and packet switching technologies. We did not include loop-related cost elements such as main distribution frames (MDFs) or line cards. For Itel, we included the DSLAM frames without line cards and MDFs.

For each switching technology, the cost is composed of two parts: initial capital investment costs and annual recurring costs. Most of the switching equipment in a CO is initial capital investment cost, while transmission links (OC-3, etc.) are considered the recurring costs. The life of telephone CO equipment can be determined in several ways. According to Internal Revenue Service documents, the product life of telephone switching equipment (Class 48.12) is 18 years.[10] Given the pace of technological change, we are doubtful that such long depreciation schedules will be sustainable in the future. Using the cost data, we have conducted sensitivity analysis of the monthly subscriber line costs with the product life varying 3 years to 20 years and minimum attractive rate of return (MARR)[11] ranging from 5% to 50%.

---

10. From IRS Publication 534, Depreciation.

FIGURE 14.5
SS7 Network Simulation Topology.

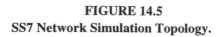

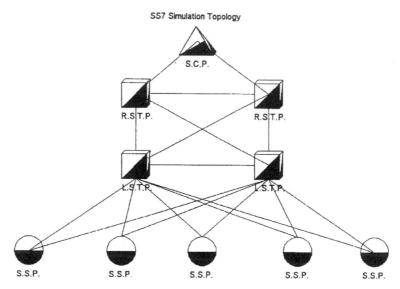

The overall results of this cost analysis are summarized in Fig. 14.8. In this figure, the solid lines are Itel costs and the broken lines are circuit switching costs, each for different values of MARR and lifetime. As the figure shows, the cost of Itel is uniformly lower than circuit switching for any value of MARR and lifetime. This result holds for all values of transmission cost and becomes stronger as transmission costs become larger.

In the following sections, we discuss the details of the cost analysis for each network technology.

Circuit Switched Costs

The simulation results indicate that a total of 14 OC-3s and 12 OC-1s would be needed for the system, resulting in a $727,000 monthly cost of trunking, assuming

---

11. MARR is frequently used in engineering economic analysis to represent the investors' expected rate of return. The MARR is defined to be the return that could be earned by investing elsewhere (opportunity cost concept). It is sometimes also called *cost of capital*. Investors use MARR, instead of interest rate, when they convert the net present value (NPV) of initial capital investment to the average recurring cost with their return over a given lifetime of the capital assests. MARR may vary depending on technology change and the riskiness of the industry investors are involved or interested in. A higher risk project would have higher MARR.

a 50% discount off of retail. The capital cost of the 5ESS switching system is around \$2.94 million[12] (switching only; Schulzrinne, 1997), resulting in a total local switching cost of \$16.03 million (initial switching investment cost). On a monthly per-line basis, the switching and trunking cost is \$2.30 (5-year life, 5% MARR), of which \$1.60 is due to the cost of trunking, showing that trunking costs dominate the overall cost of the network.

TABLE 14.5
ITU's G.114 Recommendation.

One-way delay (msec)	Description
0–150	Acceptable for most user applications.
150–400	Acceptable provided that administrations are aware of the transmission time impact on the transmission quality of user applications.
400+	Unacceptable for general network planning purposes; however it is recognized that in some exceptional cases this limit will be exceeded.

Note: ITU-T G-Series Recommendations: G.100-G.699 (available from http://www.itu.int).

Itel Costs

For the Itel network, we used the same number of COs so that the local loops would be identical between the two models. Although this might yield a suboptimal Itel design, it was necessary so that we could ignore local loop costs in our analysis. The results for Itel are the solid lines in Fig. 14.8. The net present value (NPV) cost for Itel switching and trunking was \$26.6 million, with the DSLAM contributing a high proportion of those costs when a 5-year product life and 5% MARR are assumed. To ensure acceptable quality, the network was overengineered so that each trunk was 40% utilized. With this, the monthly trunking price (at 50% off of retail) is \$286,000, and the capital cost was \$11.49 million (list price). Vendors that we spoke with remarked that a more realistic price would be a 40% to 50% discount off of list. We chose to base our analysis on the list price nonetheless, as it is the most conservative.

The monthly Itel subscriber line cost for switching and trunking is \$1.10 when the 5-year product life and 5% MARR are assumed. As with the circuit switched network, the dominant cost component is the trunking cost, which comprises \$0.63 out of the \$1.10 monthly Itel subscriber line cost for switching and trunking. Figure 14.8 (solid lines) shows the cost per line per month for varying product lifetime and MARR.

---

12. The switching cost \$2.7 per kbps for 5ESS is used.

FIGURE 14.6
Internet Telephony Central Switching Office Protocol Stack.

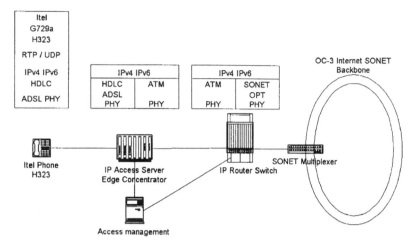

FIGURE 14.7
Packetized Voice Distribution.

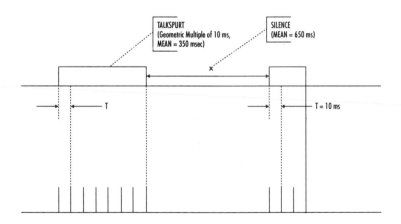

DISCUSSION AND CONCLUSIONS

The results of this analysis show that the interoffice transmission and switching costs for the Itel approach to transmission and switching are lower than for the cir-

FIGURE 14.8
Network Costs (in dollars per line per month).

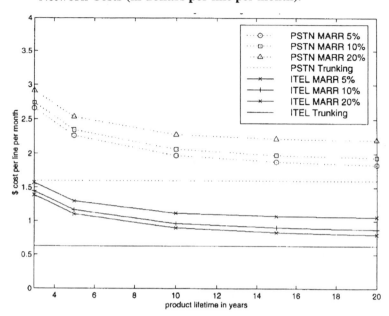

cuit switched approach for a modestly sized circuit switched network, by about $1.20 per month (or 50%)[13] using a 5% MARR and a capital equipment life of 10 years. The dominant cost factor is the cost of interoffice transmission; the compression and silence suppression enabled the use of 29% less transmission capacity[14] in Itel.[15]

This suggests that the regulatory issues raised by Itel will not go away, and that regulators will have to continue to confront them. It also suggests an imminent technology conversion for telephone companies as they continue to seek lower costs of delivering their services.

---

13. We included a Sun UltraSPARC-2 2300 at $20,000 for call processing; this may not be sufficient to handle the call processing load, so the actual capital costs for Itel may be a bit higher. Given the dominance of recurring costs in the results, we do not believe that the impact of this will be large. This is in the same price range as the Sun Netra-1 1125, which is sold to telephone companies for this purpose. We have not been able to determine how many of these systems would be necessary for a CO of the size that we have configured.

14. This cost advantage was obtained even though we used a more expensive access approach in xDSL than in the circuit switched network.

15. The Itel network required 10 OC-3s versus the 14 required for circuit switching.

Still, these results do need to be put into perspective. Generally speaking, the transmission and switching costs of an interexchange carrier (IXC) are around 22% of their total cost structure (Stuck & Weinggarten,1998), with the remainder being operations, administration, and marketing (OA&M). Some of the largest unknown costs for Itel are the OA&M costs as well as billing. These systems are highly developed for circuit switched networks and are of major importance to carriers. Finally, we did not consider operations cost differences between the two technologies, or their relative reliability, security, and so on. There is no reason to believe that these would be constant across the technologies; in fact, it is quite possible that these costs could be higher for Itel, as the systems have not yet had a chance to develop as the circuit switching systems have. However, they would have to be quite a bit higher to negate the transmission cost advantages of Itel: If the operations costs of a circuit switched carrier are 30% of the total costs, then the operations costs of Itel would have to be 34% higher for the two systems to be equivalent in cost.[16]

The costs cited here assume a 50% discount off of retail for the trunks because discounting can be expected for bulk purchases. If the carrier is facilities-based, a higher discount level would likely apply. Because the capital cost of the circuit switched network exceeds the capital cost of Itel, Itel is cheaper at all levels of discount. This is a very powerful result because it highlights the magnitude of the cost difference between the technologies.

If we add the cost of line cards and MDFs, the NPV costs of Itel and circuit switching are approximately the same at a trunking discount of 50%. Much of this is due to the use of DSL in the Itel network but not in the circuit switched network. It is quite possible that a different access technology over copper pairs for Itel might result in lower costs.

Also, we did not account for loop termination equipment or terminal devices. The total system costs must include the cost of these systems. At a high level, analog telephones are inexpensive, ranging in price from about $30 to about $150, which amounts to $0.50 to $2.50 per month over 5 years. Internet telephones are likely to be considerably more costly in the near term; future prices are likely to come down, although perhaps not to the $20 per telephone level that is available for circuit switched technology.

Another important factor is the ability to deliver integrated services. We did not consider the incremental cost of developing and deploying integrated services in

---

16. It would not be as simple to implement the compression technology in circuit switched networks, as they would have to be associated with trunks, which are application blind. Compression is optimized for voice, and will not work with fax or modem traffic, so a compression system for circuit switching would have to be able to distinguish among the various applications of a trunk, and then would have to be able to aggregate compressed traffic into 64kbps channels.

both networks. The relative cost of this (and the relative revenue opportunities based on the capabilities of the terminal device) could be an important factor in determining a carrier's choice of switching and transmission technology.

ACKNOWLEDGMENTS

We would like to thank the people at Cisco Systems, Inc., Hyperion Telecommunications, Inc., and NPT Systems, Inc., for their time and willingness to help.

REFERENCES

Bellcore. (1992, December). Signaling Transfer Point (STP) Generic Requirement, Technical reference on SS7 (Tech. Rep. No. TR–NWT–000082), Issue 4, Red Bank, NJ.

Casner, S., Frederick, R., Jacobsen, V., & Schulzrinne, H. (1996). RTP: A transport protocol for real time applications. Available ftp://ftp.isi.edu/in-notes/rfc1899.txt

Clark, D. D. (1998). A taxonomy of Internet telephony applications. In J. K. MacKie-Mason & D. Waterman (Eds.), *Telephony, the Internet, and the media: Selected papers from the 1997 Telecommunications Policy Research Conference* (pp. 157-176). Mahwah, NJ: Lawrence Erlbaum Associates.

Frieden, R. (1996, October). *Dialing for dollars: Will the FCC regulate Internet telephony?* Paper presented at the 1996 Telecommunications Policy Research Conference, Solomons, MD.

Kostas, T. J., Borella, M. S., Sidhu, I., Schuster, G. M., Grabiec, J., & Mahler, J. (1998, January–February). Real-time voice over packet-switched networks. *IEEE Network*, pp. 18–27.

Kumar, V. P., Lakshiman., & Stiliadis, D. (1998, May). Beyond best effort: Router architectures for the differentiated services of tomorrow's Internet. *IEEE Communications Magazine*, pp. 152–164.

Low, C. (1997, June). Integrating communication services. *IEEE Communications Magazine*, pp. 164–169.

Manchester, J., Anderson, J., Doshi, B., & Dravida, S. (1998, May). IP over SONET. *IEEE Communications Magazine*, pp. 136–142.

Mueller, M. L. (1997). *Universal service.* Cambridge, MA: MIT Press.

Schulzrinne, H. (1994). RTP profile for audio and video conferences with minimal control. Available ftp://ftp.isi.edu/innotes/rfc1890.txt

Schulzrinne, H. (1997). Re-engineering the telephone system. Available www.cs.columbia.edu/~hgs/papers/Schu9704a-Reengineering.ps.gz

Stuck, B., & Weinggarten, M. (1998, August). Can carriers make money on IP telephony? *Business Communications Review*, pp. 39–54.

15

An IP-Based Local Access Network: Economic and Public Policy Analysis

Daniel Fryxell
Marvin Sirbu
Kanchana Wanichkorn
Carnegie Mellon University

This chapter investigates the costs a local exchange carrier (LEC) would incur to implement an Internet protocol (IP)-based local access network providing both voice and Internet access service. The architecture employs end-to-end IP technology over asymmetric digital subscriber lines (ADSL) in the local loop. A cost model of key network elements estimates the forward-looking cost of this integrated network. The cost of the proposed architecture is compared with the cost of a traditional network solution for voice and Internet access using the public switched telephone network (PSTN). In this baseline architecture, the PSTN is used for Internet access and Internet users are provided with an extra analog line for a voice-grade modem. Because multiple voice conversations can be carried over a single ADSL loop, the IP architecture dominates only where loop savings for multiline subscribers exceed the additional electronics costs of ADSL. A number of technical and policy issues regarding the deployment of the proposed network are also discussed in the last part of this chapter.

INTRODUCTION AND MOTIVATION

It has taken nearly 30 years for the historically analog telephone network to be transformed into today's end-to-end digital circuit switched network. Today we are on the verge of a second revolution in telephone technology that will be every bit as radical as the conversion from analog to digital: from circuit switched to packet switched technology for the carriage of voice as well as data traffic (Turner, 1986). The development of technology for carrying voice traffic over IP networks has already resulted in the widespread introduction of packet switched technology in the long-distance network at significantly reduced tariffs (Okubo et al., 1997; Thom, 1996). Avoided access charges account for part of these savings along with the use of advanced compression techniques to reduce the bit rate of a call to as little as 8 Kbps from the standard pulse code modulation (PCM) rate of 64 Kbps. Numerous firms sell gateway products for conversion between circuit switching and voice over Internet protocol (VOIP). Several firms have introduced IP-based private branch exchange (PBX) products as well. As with the conversion from analog to digital, the last part of the network to switch from circuit to packet technology will be the subscriber loop and the residential customer.

The shift from circuit switched to packet switched networks is based on several converging trends. The most important is the emergence of data as the dominant form of traffic on the nation's telecommunications networks. Various commentators have set the crossover date between 1998 and 2002. Virtually all agree, however, that with voice traffic growing at only 6% to 8% per year as data traffic grows 30% to 40% per year, it will take only a few years after the crossover for voice to represent but a small proportion of total telecommunications traffic. Thus, in addition to any cost or functionality benefits that may be available by carrying voice over IP networks, carriers will find it irresistible to combine voice and data on a single network to simplify network planning and administration.

In the balance of this chapter, we present an economic analysis of an IP-based local access network, providing both voice and data services. We compare the cost of this integrated packet network to the cost of a traditional PSTN with a second phone line used to call an ISP. The next section describes the architecture of the proposed network. The cost model and its results are presented following that. We then discuss a number of technical and policy issues regarding the deployment of this integrated packet network. The final section of the chapter provides a conclusion.

PROPOSED NETWORK

The proposed architecture envisions the concept of end-to-end IP technology with ADSL in the local loop to provide complete integration of voice and Internet access (see Fig. 15.1).

ADSL is a digital technology for the local loop that utilizes the existing copper infrastructure to affordably deliver broadband data rates to customer premises. This technology is attractive to incumbent LECs since it makes use of their ubiquitous copper wire plant. ADSL delivers downstream (from the central office to the customer premises) payloads of up to 6 Mbps and upstream payloads of up to 640 Kbps on 24-gauge loops of up to 12,000 feet. For distances up to 18,000 feet, the data rates decrease to 1.5 Mbps downstream and 160 Kbps upstream. This technology provides sufficient bandwidth for multiple simultaneous VOIP conversations.

FIGURE 15.1
IP-based Network.

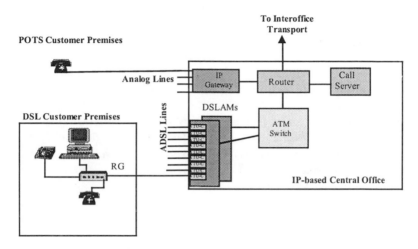

At the customer premises, an ADSL line terminates at a remote ADSL transceiver unit (ATU–R), which is part of the residential gateway (RG) (see Holliday, 1997). The RG is the demarcation point between the customer premises equipment (CPE) and the subscriber loop network. During the initial transition to a totally IP network, RGs may contain a VOIP gateway function that converts between analog voice signals and VOIP packets. In this scenario, RGs are equipped with RJ11 ports for analog telephone handsets and an RJ-45 port to connect Ethernet-based equipment, such as a personal computer. In the future, the RG may be the gateway for a home area network (HAN)—a local area network (LAN) optimized for the home. In the future, handsets may send packets over the HAN as opposed to analog voice. Several technologies have been suggested for this HAN network (Freed, 1998). The Home Phoneline Networking Alliance (HomePNA) is developing Ethernet technology that can run over today's in-home wiring with all its multiple taps and splices (Niccolai, 1998). The Home Radio Frequency (RF) alliance is developing standards for a wireless solution (Simple Wireless Access Protocol; SWAP) to the home networking problem (Ohr & Boyd-Merritt, 1998). IEEE 1364 ("Firewire") based standards for high-speed home networking could support video as well as voice and data services. All of these home networks provide enough bandwidth for VOIP traffic.

Today's telephone switch consists of three main components: line cards, which terminate subscriber loops and interoffice trunks; the switch fabric, which connects traffic between lines or to trunks; and the common control computer, which provides call setup and custom calling and supports advanced intelligent network services.

The IP-based central office (CO) presented in Fig. 15.1 provides the same functionality in a radically different way. At the CO, ADSL lines terminate in CO transceiver units (ATU–Cs) that are housed by digital subscriber line access multiplexers (DSLAMs). DSLAMs statistically concentrate the typically bursty individual streams into high-bandwidth links. A router, which is connected to the DSLAMs via an asynchronous transfer mode (ATM) switch, provides the logical switching function.

Currently, the most popular architecture for ADSL access networks is based on an ATM layer running over the DSL physical layer (ADSL Forum, 1997; Hawley, 1997; Humphrey & Freeman, 1997). In this model, communication channels between RGs and the router are ATM virtual circuits managed centrally at the CO. The ATM switch provides the ATM user network interface and aggregates traffic coming from different DSLAMs. Running over ATM, the IP layer provides a common communications protocol for the entire network (from CPE to interoffice transport).

Common control functions are provided by server programs running on one or more stand-alone processors. These elements provide the necessary signaling and call control functions, such as H.323 gatekeeper functionality, domain name translation, mapping E.164 numbers to IP addresses, and so on. Finally, an IP gateway, which converts analog voice to IP packets and vice versa, provides connection to analog voice lines. This device gives customers with less demanding communication needs the alternative to simply choose analog voice lines. The functionality required for the gateway can be performed by a Class 5 telephone switch with an IP interface on the trunk side.

Competitive subscriber loop networks provided by cable or wireless carriers could be similarly configured to provide integrated IP service.

COST MODEL

In this section, we develop a cost model for the IP-based broadband architecture just presented. We compare the cost of the proposed architecture with the cost of a traditional network solution for voice and Internet access using the PSTN. In the PSTN-based architecture, Internet access is provided by dial-in connections that make use of voice-grade modems for data transfer. Currently, the majority of Internet subscribers connect their modems via the same lines they use for voice service. However, in this chapter we assume that Internet subscribers purchase an additional phone line for the modem. We make this assumption to have some parallelism with the IP/ADSL architecture, where voice lines are not blocked by Internet access.

It is important to note that voice and Internet access services do not necessarily have the same quality in both architectures. The speed of Internet access and the "always-on" availability, for instance, differ considerably between the two cases.

Only costs that are directly associated with the network infrastructure are included in the model; that is, capital carrying costs and maintenance expenses. Expenses for operations and support, as well as marketing, general and administrative (G&A), and other costs of running a company, are not included.

Throughout the cost model we assume a forward-looking perspective. This means that we determine the costs for a greenfield network built today with current technology. The fact that there are sunk costs in legacy LEC networks is not considered in the cost model.

General Layout

In this chapter we analyze architectures for a network of an LEC within a local access and transport area (LATA) that provides voice and Internet access service. The total area covered by the network is divided into elementary areas served by a local CO. Customer premises are connected to COs by local loop lines (Fig. 15.2). COs are connected to each other by interoffice transport facilities (Fig. 15.3). COs are the interconnection point for lines in the same elementary area and the gateway to the interoffice facilities, which provide connection to locations outside the area served by the CO.

FIGURE 15.2
Central Office Area.

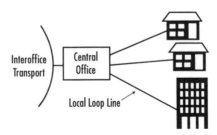

All COs are assumed connected to a single synchronous optical network (SONET) fiber ring for interoffice transport within a LATA. This configuration is not necessarily the optimal solution for any particular LATA. However, the one-ring architecture assumption provides a simple model to estimate the requirements of the interoffice transport infrastructure.

FIGURE 15.3
Interoffice Transport.

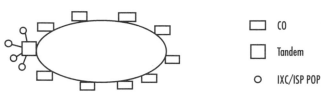

Interoffice transport facilities also include a tandem office that routes indirect channels between COs and serves as the gateway to the points of presence (POPs) of interexchange carriers (IXCs) and ISPs. The connections between the tandem and the IXC or ISP POPs are modeled as point-to-point fiber links. Although the SONET architecture allows the provisioning of direct channels between any pair of COs, routing some connections via the tandem allows a more efficient allocation of bandwidth.

We assume that interLATA voice traffic and Internet access traffic are delivered to generic IXC/ISP POPs that accept both voice and Internet access traffic. In current voice networks there are many small ISPs connected to COs instead of large ISPs connected to tandems. However, the configuration with ISPs at the tandem is more appropriate for the IP/ADSL architecture, where there is no distinction between voice and data. For comparison, we assume the same configuration for the PSTN-based architecture. By assuming that customers connect directly to a few backbone POPs, this PSTN-based architecture internalizes the costs of the dedicated lines that ISPs currently use to connect to Internet backbones, as these costs are automatically internalized in an IP/ADSL configuration.

Characterization of the Locations Served by the Network

The architectures analyzed here are suitable to provide Internet access to households and small businesses. Medium and large corporations have access to higher capacity and more expensive networking solutions. We adopt the simplified assumption that all the customers were either residential or small businesses. In the PSTN-based architecture, the number of voice lines determines the number of loop lines for voice and the number of dial-in modem ports determines the number of loop lines to an ISP for Internet access (Table 15.1). This means that two loop lines are needed per household and four loop lines are needed per small business location. In the IP-based architecture, voice and Internet access traffic are carried over a single ADSL line; thus, only one loop is needed per household or small business location.

TABLE 15.1
Average Number of Lines Per Location.

	Household	Small business location
Number of voice lines per location	1	3
Number of dial-in lines per location (PSTN-based architecture)	1	1

Traffic

We assume 50,000 voice lines per CO, of which 65% are residential lines and 35% are small business lines. The LATA has 30 equivalent COs with a total of 1.5 million voice lines. The total number of locations with Internet access is determined by multiplying the number of locations by the Internet penetration parameter.[1]

VOICE TRAFFIC

Busy hour call attempts (BHCA) per line, call completion ratio (calls completed over calls attempted), intraCO fraction, and tandem-routed interLATA fractions (Table 15.2) are assumed as suggested by Hatfield (1997). The tandem-routed intraLATA fraction is assumed as an average of the values for local and intraLATA suggested by Hatfield (1997).

TABLE 15.2
Voice Traffic.

	Residential	Business
Busy hour call attempts (BHCA)	1.3	3
Call completion ratio (calls completed/calls attempted)	70%	70%
Call holding time (minutes)	3	3
Fraction of intraCO calls	35%	35%
Fraction of intraLATA calls	50%	37%
Fraction of interLATA calls	15%	28%
Tandem-routed fraction of intraLATA	10%	
Tandem-routed fraction of interLATA	20%	
Data rate of an IP voice channel (Kbps)	64	

---

1. We use a default of 100%, implying all household and business locations in the serving area purchase Internet access.

The fractions of intraLATA and interLATA traffic were estimated based on the values of dial equipment minutes (DEMs) and call completion reported by the Federal Communications Commission (FCC) for the state of Pennsylvania and the values suggested by Hatfield (1997) for the business/residential DEMs ratios for local, intraLATA, and interLATA traffic.

Call holding time (assumed constant) was determined as the value that made the busy hour CCS^2 (BHCCS) derived from the BHCA and the call completion fraction the same as the BHCCS derived from the DEMs, assuming busy hour traffic concentration factor of one tenth of total DEMs divided by 270 business days.

Voice traffic in the IP/ADSL architecture is assumed 64 Kbps per voice channel (ITU-T Recommendation G.711). VOIP standards include codecs such as G.723.1 operating at bit rates as low as 5 to 6 Kbps, but we chose to err on the side of higher bit rate voice.[3]

Internet Access Traffic

For the dial-in architecture, Internet access traffic is expressed in CCS because the network is circuit switching. However, for the more flexible packet-based IP/ADSL architecture this traffic is expressed as a data rate.

Traffic assumptions for the dial-in architecture are based on a study that monitored ISP lines serving mainly residential subscribers (Morgan, 1998). ISP lines have a peak traffic around 27 CCS. This value goes down to about 20 CCS during the typical busy hours of voice networks. According to Leida (1998), ISPs size their modem access servers to have about 10 residential users per modem. Based on these values, we can estimate the busy hour probability that a given residential customer is connected to his or her ISP as

$$10\% \bullet \frac{27CCS}{36CCS} = 7.5\%$$

with an expected usage during the peak hour of 2.7 CCS; and a probability of

$$10\% \bullet \frac{20CCS}{36CCS} = 5.56\%$$

during the network busy hour (or 2 CCS per subscriber).

For the PSTN-based architecture studied here, probabilities are likely to be higher than these values because Internet subscribers have an extra line for the modem. To reflect this factor, we increased the activity rates by 30%. The activity rate for small business is assumed twice that of residential subscribers, with the business peak occurring at the voice busy hour.

---

2. Hundreds of call seconds. One circuit occupied during 1 hour is equivalent to 36 CCS.
3. As Fig. 15.10 shows, costs are not sensitive to this assumption.

We assume that the activity rates derived for dial-in connections stay the same for the ADSL architecture. The average bandwidth usage when active is assumed twice the average bandwidth for modem users suggested by Leida (1998) recognizing that with faster DSL links users will likely download more bits per active hour. The uplink bandwidth is assumed 10% of the downlink bandwidth (Table 15.3).

TABLE 15.3
Internet Access Traffic.

	Residential	Business
Busy hour activity rate	7.4%	20%
Average downlink data rate in the IP-based architecture (Kbps)	10%	
Uplink data rate as a fraction of downlink data rate	10%	

Local Loop

We assume local loops are all copper from the customer premises to the CO. Because ATU–Cs must be located where the copper ends, this is a necessary assumption to have all the DSLAMs located in the CO. The inclusion of a digital loop carrier (DLC), which concentrates copper pairs at a remote terminal with a fiber feeder from the CO, would introduce extra complexity into the model. In that case, because copper lines would not reach the CO, DSLAMs would have to be moved from the CO to the remote terminals. Our assumption is consistent with the fact that ADSL technology is first being deployed in areas where loops are short and DLC systems have not been deployed. Estimates for local loop investment are derived using the Hatfield Model for an area (Washington, DC) where little use of DLC is required[4] and thus are consistent with the assumption that all the copper lines terminate at the CO.

Loop costs include loop cable, installation, and infrastructure (conduits, poles, cable protection, etc.). Because the last two components account for a large part of the loop cost and are not very sensitive to the number of lines per location, the marginal cost of adding extra lines is considerably lower than the cost of providing the first line. Cost per location passed and incremental cost per line were estimated as the coefficients of a linear regression on the loop cost per location versus the number of lines per location calculated by modifying the Hatfield Model (Table 15.4).

---

4. The results of the Hatfield Model for Washington, DC show DLC appropriate for only 16% of lines. Criticisms of the loop costs calculated by the Hatfield Model (Schechter, 1998) do not apply in the urban areas we are examining.

TABLE 15.4
Local Loop Lines.

	Investment
Fixed cost per location passed	$413
Incremental cost per line	$205

CO

IP-BASED CO

The IP-based CO configuration is as discussed in Proposed Network previously. The cost model includes the elements shown in Figure 15.1 plus a main distribution frame (MDF) and add drop multiplexers (ADMs). The MDF is the endpoint for the copper lines of the local loop. This element provides electrical protection and is the interface between the loop lines and the in-office equipment. The MDF investment is computed based on a cost per line, for which we use the value suggested by Hatfield (1997).

The values for cost and capacity of DSLAMs were derived from equipment currently available in the market. Prices of ADSL equipment still show a large variation and rapid price reductions, which are typical for new technologies in the initial phase of deployment. (See later discussion in Results and Sensitivity Analyses for a sensitivity analysis on this parameter.)

TABLE 15.5
IP-based Central Office.

Main distribution frame		ATM switch	
Busy hour activity rate	$17.5	Common equipment	$28,000
DSLAM		OC-3 port	$2,000
Cost per line	$400	DS-3 port	$1,000
Installation cost	10%	Installation cost	10%
Maximum allowable fill	98%	**Router**	
ATU-Cs per DSLAM	200	Common equipment	$60,000
IP gateway		OC-3 port	$10,000
Cost per line	$100	Installation cost	10%
Installation cost	10%	**Transmission equipment**	
Maximum allowable fill	98%	OC-48 ADM, installed	$50,000

IP gateways with the level of concentration required for a CO are just now emerging (Table 15.5). Commercial products currently targeting the corporate market can only terminate a small number of voice lines. The assumption for the cost of this element, which is only a rough estimate, is based on prices of existing products taking into consideration economies of scale.

The ATM switch and the router are sized based on the number and type of connections such that their forwarding capacities are not a bottleneck. Links between DSLAMs and the ATM switch are at the DS-3 rate, while all the links to the router are at the OC-3 rate. ATM switch and router costs were derived from current list prices[5] with a 30% discount.

The call server in the IP-based architecture performs functions comparable to the common control of a circuit switch in the PSTN. For this reason, the server cost was assumed as the cost of common control for an equivalent PSTN switch. This value was estimated as 70% of the switch common equipment (common control plus switch fabric), which was determined as total cost of the switch (computed as in Hatfield, 1997) less the cost of line and trunk cards.[6]

ADMs connect the SONET ring, which supports OC-48 circuits,[7] to the transmission equipment inside the CO. Per each OC-48 circuit to which the CO is connected, there is one ADM that extracts or inserts OC-1 channels from or into the OC-48 circuit. ADMs are connected to the router by OC-3[8] links. OC-1 channels in OC-48 circuits can be individually allocated to establish transport links between pairs of COs, COs and the tandem switch, or COs and IXC/ISP's POPs. Direct channels, either between pairs of COs or COs and POPs, are established only if there is enough traffic to justify them. Otherwise, traffic follows an indirect path via the tandem router. Nondirect traffic shares the channels that are allocated for transport between the tandem router and COs or POPs.

PSTN-BASED CO

The basic configuration of the PSTN network used in this chapter is derived from the Hatfield Model (Hatfield, 1997). The cost model includes the elements shown in Figure 15.4.

The MDF and the ADM perform basically the same function as in the IP-based architecture. In this architecture, the basic transport channels, which are extracted or inserted from or into the OC-48[9] interoffice circuits by ADMs, are DS-3s.

---

5. For list prices, see http://www.networkcomputing.com/.

6. For this purpose, the cost per line card is assumed to be $60.

7. An OC-48 circuit carries 48 OC-1 channels at 51,840 Mbps each for an aggregate data rate of 2,488,320 Mbps.

8. An OC-3 circuit carries three OC-1 channels.

9. An OC-48 circuit carries 48 DS-3 channels, which is equivalent to 1,344 DS-1s or 32,256 DS-0 voice channels.

FIGURE 15.4
PSTN-based Central Office.

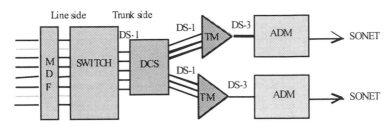

A narrowband circuit switch is at the core of this architecture. This PSTN switch establishes intraoffice connections between pairs of loop lines and interoffice connections between loop lines and interoffice trunks. The investment in switching equipment is determined by adding the investment in line cards, common equipment (including switch fabric and common control), and trunk ports using costs found in Hatfield (1997). The sum of the first two terms is given by the product of the number of lines by the switch cost per line, which is estimated as a logarithmic function of the number of lines given by

$$\text{switch cost per line} = A_{switch} \ln \left(\text{number of lines}\right) + B_{switch}$$

Grooming equipment for the CO includes digital cross-connect systems (DCSs) and terminal multiplexers (TMs). These elements are sized according to the interoffice traffic, which is computed based on the assumptions for traffic generation and routing. DCSs cross-connect DS-1 trunks between the switch and TMs. TMs, which provide an interface to the switch at the DS-1 level, multiplex DS-1 circuits into the DS-3 channels used in interoffice transport (Table 15.6).

TABLE 15.6
PSTN-based Central Office.

Main distribution frame		Switch	
Cost per line	$17.5	A_{switch}	-14.9
Grooming equipment		B_{switch}	242.7
DCS, per DS-3, installed	$30,000	Maximum allowable fill (line side)	98%
OC-3/DS-1 TM, installed	$26,000	Cost per trunk port	$100
OC-48 ADM, installed	$50,000	Maximum trunk occupancy (CCS)	27.5
Maximum allowable fill	80%	Switch installation cost	10%

Interoffice Cable

The total length of an interoffice ring is determined by summing the lengths of the ring that cross each CO area. To estimate the dimensions of each CO area, we assume that these areas are square and there is a uniform line density throughout the LATA of 5,000 lines per square mile. The distance that the ring crosses within each CO area is the side of the square multiplied by a factor of 1.5, which accounts for the fact that the route of the cable is not a straight line.

The tandem office is assumed connected to five different IXC/ISP's POPs by links that are half a mile long.

The number of fiber strands per cable is a multiple of 24. Each circuit in the ring requires two strands and each point-to-point circuit, such as the ones between the tandem and the IXC/ISP's POPs, requires four strands. We assumed no use of dense wavelength division multiplexing (DWDM).

The costs of fiber cable and cable infrastructure (protection, poles, conduits, and installation) shown in Table 15.7 are the aggregate values of all the subelements into which they can be decomposed (for a more detailed description see Fryxell, 1998).

TABLE 15.7
Interoffice Transport.

24-fiber interoffice cable (per foot)	$3.5
24-fiber increment (per foot)	$1.2
Interoffice cable infrastructure (per foot)	$1.1

Tandem Office

At the core of the tandem office is a router (or a switch for the PSTN-based architecture) that provides the logical switching functionality to handle nondirect intraLATA and interLATA traffic. The tandem office has ADMs that extract or insert basic OC-1 transport channels from or into the SONET ring. These channels include direct channels between COs and IXC/ISP's POPs and channels between COs and the tandem router (or switch). Direct channels are physically cross-connected from the ADMs to the links connecting the tandem to the IXC/ISP's POPs. Channels between COs and the tandem router (or switch) carry intraLATA and interLATA traffic that, for a more efficient allocation of channels, is indirectly routed via the tandem router (or switch).

The cable links between the tandem and the IXC/ISP's POPs use OC-3 circuits. It is not cost-effective to multiplex OC-3 circuits into a higher capacity OC-48 circuit because these links are very short. Savings in fiber would be outweighed by the extra cost in ADMs. Because these links carry both direct and tandem-routed traffic, some of the OC-3 circuits come directly from the ADMs, while the rest come from the tandem router (or switch).

IP-BASED TANDEM OFFICE

In the IP approach, the architecture for the tandem office consists of a router and ADMs. All the links to the router are at the OC-3 level (Table 15.8).

TABLE 15.8
IP-based Tandem Office.

Router	
Common equipment	$60,000
OC-3 port	$10,000
Installation cost	10%
Transmission equipment	
OC-48 ADM, installed	$50,000

PSTN-BASED TANDEM OFFICE

In the PSTN-based architecture, the tandem office consists of a circuit switch, ADMs, and TMs. TMs provide interfaces between the switch and the ADMs, and the switch and the links to IXC/ISP's POPs.

The tandem switch investment was determined as in Hatfield (1997). This investment is the sum of two components: an investment in trunk ports based on a fixed cost per trunk and an investment in common equipment that scales linearly with the number of trunk ports (Table 15.9).

TABLE 15.9
PSTN-based Tandem Office.

Switch		Grooming equipment	
Cost per trunk	$100	OC-3/DS-1 TM, installed	$26,000
Minimum common equipment cost (0 trunks)	$500,000	OC-48 ADM, installed	$50,000
Maximum common equipment cost (100,000 trunks)	$1,000,000	Maximum allowable fill	80%
Switch installation cost	10%		
Maximum allowable fill	80%		

Other Equipment

CPE

In the IP-based architecture, the RG is the only CPE included in the cost model. The cost for this element was determined based on prices of small ISDN routers with analog telephone ports. In the PSTN-based approach, CPE for Internet access is a 56 Kbps modem. The cost of this element was derived from typical prices paid in retail stores (Table 15.10).

TABLE 15.10
Customer Premises Equipment.

IP-based architecture	$500
PSTN-based architecture	$100

Analog telephone handsets are also necessary for both architectures, but this low-cost element is not included in the cost model.

IN-BUILDING WIRING

As we mentioned before, the cost model for the IP architecture excludes HAN and CPE costs except for the RG. With SWAP and/or the HomePNA these costs will be minimal. By contrast retrofitting a home for LAN service over CAT5 cabling can easily cost over $1,000 (Bill, 1998).

ISP EQUIPMENT

Because the network analyzed in this chapter is a local network, equipment located at IXC/ISP's POPs should be out of its scope. However, the IP-based and PSTN-based architectures differ substantially regarding the way interLATA traffic is delivered to ISP POPs. While in the former case Internet access traffic is already delivered in IP packets over high-capacity circuits, in the latter case traffic is delivered as modem signals over voice lines. For a fair comparison, we include the elements that make the conversion from analog signals over voice lines to data packets over high-speed data lines. These elements are remote access servers, which are banks of modems and access routers, and the TMs that provide an interface to these servers at the DS-1 level.

The investment per modem for remote access servers was derived from list prices with a discount of 30%. The investment per DS-3 TM is estimated as one third of the investment per OC-3 TM (Table 15.11).

TABLE 15.11

ISP Equipment.

Terminal multiplexes		Remote access servers	
Investment per DS-1/DS-3 TM	$8,700	Investment per modem	$250
Maximum allowable fill	80%	Maximum allowable fill	98%

Capital Carrying Costs and Maintenance Expenses

Capital carrying costs are determined based on equipment service lives. For the LEC part of the network, with the exception of switching equipment, we use the lives adjusted for net salvage value proposed in Hatfield (1997).[10] Reflecting the current rate of change in the industry, a value of 10 years, rather than 16, for switching equipment seems more appropriate for a forward-looking perspective. Moreover, this value is consistent with the life of transmission equipment (Table 15.12).

TABLE 15.12

Lives and Maintenance Expenses for Network Equipment.

	Adjusted projection life	Expense factor
Switching	10	2.7%
Digital circuit equipment	10	1.5%
Interoffice fiber cable	23	7.2%
Interoffice cable infrastructure	24	3.6%
Local loops	28	3.7%
CPE equipment	10	1.5%
ISP equipment (modems and TMs)	10	1.5%

The annual maintenance expense for each of the network elements is determined as the initial cost of the element times its respective expense factor. Expense factors were derived from historic expense ratios for Bell Operating Companies (BOCs),[11] as reported on their balance sheets and expense account information.[12] As suggested in Hatfield (1997), expense factors for switching and transmission

---

10. These values are based on the average projection lives as determined by the-three way meetings (FCC, State Commission, and ILEC) for the BOCs and SNET.

11. The numbers were calculated as the average of the values for the states of Pennsylvania, California, Maryland, and the District of Columbia.

12. This information may be found on the ARMIS report, which is in the appendix of the Hatfield Model.

equipment are forward-looking values derived from a New England Telephone cost study.[13]

Service lives for CPE and ISP equipment are assumed 10 years (the same as for LECs' switching and transmission equipment). In reality, ISPs use much shorter periods—on the order of 3 or 4 years—to depreciate their investment in communications equipment. However, we chose this value in order to have only one life for all the transmission equipment in the network. The effect is to bias the cost of the PSTN solution downward, making our comparison more conservative. In any case, ISP equipment is a small part of PSTN configuration costs.

The rate of return for the LEC part of the network was determined by weighting the return on equity and the return on debt by their respective fractions, as given by Hatfield (1997). The rate of return for ISP equipment is the value suggested by Leida (1998). (See Table 15.13.)

TABLE 15.13
Rate of Return on Investment.

Network equipment	10%
CPE	15%
ISP equipment	20%

Results and Sensitivity Analyses

MONTHLY COST PER CUSTOMER LOCATION

Figure 15.5 compares monthly cost per customer location between the PSTN and the IP/ADSL architectures. Total cost of the IP/ADSL architecture is higher than for the PSTN architecture for residential customers but lower for business customers. For both residential and business customers local loop costs are lower in the IP/ADSL architecture because fewer loops are required. However, for residential customers these savings are not large enough to offset the higher investment in CPE and CO equipment. On the other hand, for business customers the savings due to fewer loops outweigh the higher investment in CPE and CO equipment, resulting in a lower total cost for the IP/ADSL architecture.

Figure 15.6 shows the cost breakdown of CO equipment and interoffice transport for the IP-based architecture. It is apparent that DSLAM costs dominate.

---

13. New England Telephone, 1993 New Hampshire Incremental Cost Study, Provided in compliance with New Hampshire Public Utility Commission Order Number 20, 082, Docket 89–010/85–185, March 11.

FIGURE 15.5
Average Monthly Cost Per Customer Location.

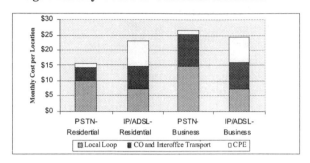

FIGURE 15.6
IP-based CO and Interoffice Transport Cost Breakdown.

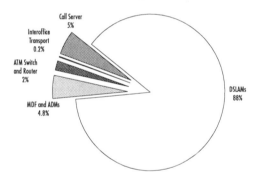

SENSITIVITY ANALYSIS OF RESIDENTIAL GATEWAY AND DSLAM COSTS

Figure 15.7 presents a two-way sensitivity analysis of the combined effect of variations in RG and DSLAM costs on the monthly cost: The PSTN is preferred when RG and DSLAM costs are high and IP/ADSL is preferred when the opposite is true. At the default values of $500 for the RG cost and $400 for DSLAM cost, IP/ADSL architecture is preferred for business customers but PSTN is preferred for residential customers. According to the analysis shown here, we can expect that the IP/ADSL architecture will be cost-effective for residential customers if RG and DSLAM costs continue to decline.

SENSITIVITY ANALYSIS OF THE NUMBER OF VOICE LINES PER LOCATION

Figure 15.8 shows monthly cost per location as a function of the number of voice lines: Break-even is at 3.0 lines for residential customers and 2.6 lines for business customers. For our default values of one line per home and three lines per business location, PSTN is preferred for residential customers and IP/ADSL is preferred for business customers.

In the PSTN scenario, costs increase proportionally with an increase in the number of voice lines because more loops are required to the customer premises. On the other hand, costs are insensitive in the IP/ADSL case because a single high-speed digital subscriber line can support multiple telephone conversations simultaneously, thereby obviating the need for additional loops to provide additional voice channels.

FIGURE 15.7
Preference Regions As a Function of Residential Gateway and DSLAM Costs.

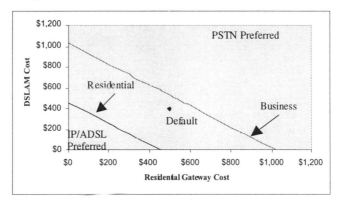

SENSITIVITY ANALYSIS OF BANDWIDTH USAGE

Figure 15.9 shows cost per location as a function of average bandwidth usage for Internet access. Results are very clear that costs are basically fixed and barely change with bandwidth usage. However, it is important to note that these costs are for the local infrastructure. Internet backbone costs are certainly more sensitive to this parameter. The PSTN architecture is circuit switching and therefore bandwidth usage in the local exchange area is not a question. Dial-in connections always use the entire circuit independently of the actual data transferred. Costs are similarly insensitive to the bandwidth use per voice channel as shown in Figure 15.10.

SENSITIVITY ANALYSIS OF INTERNET PENETRATION RATE

The results shown are based on the assumption that all homes and business locations in the serving area require both voice and data services. In reality, some may need only basic voice service. Figure 15.11 illustrates the changes in requirements if not all residential and business customers subscribe for an Internet access service. In the PSTN case, not subscribing to the Internet means dropping the extra dial-in line; in the IP/ADSL case it means using analog voice lines terminated at the IP gateway (see Figure 15.1).

As stated earlier, IP/ADSL is not cost-effective for residential customers at our projected equipment costs. Even though it requires fewer loops to the homes, savings in loop costs are not high enough to justify the higher investment in CPE and CO equipment. IP/ADSL is cost-effective for business customers because savings in loop costs more than offset higher CPE and CO equipment costs. Figures 15.12 (residential) and Figure 15.13 (business) confirm that these findings are insensitive to assumed rates of Internet penetration.

FIGURE 15.8
Monthly Cost As a Function of Number of Voice Lines Per Location.

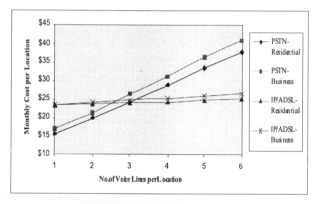

FIGURE 15.9
Monthly Cost Per Location As a Function of Internet Bandwidth Usage.

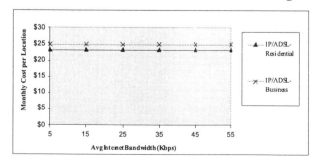

Limitations

Our analysis here has focused on installed first capital costs. We have not explored the extent to which operations support costs for this new network design may be greater or less than for current PSTN technology. Given the maturity of the latter, operations support systems are presently more highly developed. Thus in the near term, we might expect operations costs for an IP/ADSL network to be higher than for the PSTN configuration.

OTHER ISSUES

The transition from circuit switching to VOIP raises a number of technical and policy issues that will have to be addressed before the conversion can become widespread.

FIGURE 15.10
Monthly Cost Per Location As a Function of BW/voice Channels.

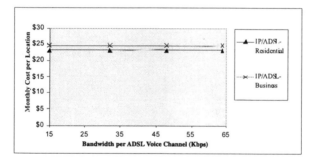

FIGURE 15.11
Customer Characteristics.

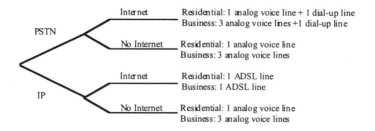

Technical Issues

Although it is possible to put together proof-of-concept systems using currently available technology, widespread diffusion will require resolution of a number of still-unresolved issues.

INTERNET QUALITY OF SERVICE

Today's public Internet provides only a "best-effort" level of service. At peak hours, packet delay and loss rates, particularly at exchange points between carriers, can be significant. There are two approaches to dealing with the quality of service problem. Many of today's VOIP service providers carry the traffic over private internets where loads are managed to ensure minimal packet delays. At the same time, new protocols to provide differentiated service over the public Internet are being discussed within Internet standards bodies. Agreement on standards and their deployment among the diverse carriers comprising the Internet will take several years. Further, introduction of differentiated services requires differentiated prices. New pricing and billing mechanisms will be needed to accompany the introduction of differentiated quality of service (Clark, 1998). Finally, the reliability of today's Internet—as measured by the frequency and severity of service outages—still does not meet the standards of today's PSTN.

OPERATIONS SUPPORT

As the faltering roll-out of ISDN demonstrated, the development of operational support systems—including network management, ordering, provisioning, and billing—can stall the deployment of a technology that works perfectly on a small scale. Because so many elements of the network will change, the new operations systems that will be required are many.

DIRECTORY SERVICES

Voice telephone users today are identified using the ITU's E.164 standard for telephone numbers. VOIP endpoints use Internet addresses. In order that callers be able to identify callees using a uniform name space, most VOIP phones will be assigned an E.164 number as an alias. This then poses the need for a directory service that maps E.164 numbers into the appropriate Internet address. For VOIP users who wish to reach callees on the legacy PSTN, there must be an equivalent service for mapping the callee's E.164 number into the IP address of the most appropriate gateway. Standards for these directory services have yet to be developed. Moreover, as the current controversy over domain names demonstrates, developing a worldwide infrastructure for name space management is no easy task.

FIGURE 15.12
Monthly Cost Per Home As a Function of Internet Penetration.

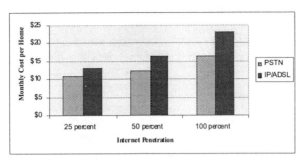

FIGURE 15.13
Monthly Cost Per Business As a Function of Internet Penetration.

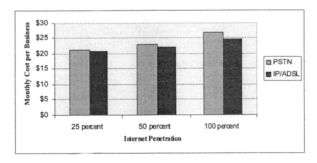

Policy Issues

POWER AND EMERGENCY SERVICE

Today's telephone service is powered from the CO and continues to function when the electric utility fails. However, RGs and VOIP handsets are likely to draw considerably more power than today's telephones, requiring connection to household power outlets. Thus, during an electric utility outage, the telephone might not operate. Potential solutions to this problem include backup battery power integrated into the RG, or substantial reductions in the power consumption of both a VOIP handset and gateway that would allow both to be powered from the CO. It is not clear whether HANs based on the HomePNA standards will support powering through the home wiring. For the case where the HAN is based on wireless technology, such as SWAP, we can assume that the handsets are already battery powered; the need to power the DSL remote and the SWAP base station remains, however.

COMPETITION IN VERTICAL SERVICES

In the PSTN, the local telephone CO integrates the line card, switching fabric, and common control processor into a monolithic telephone switch. In the VOIP world, switching and common control functions may be physically and even administratively separated. Thus, it is perfectly feasible for a subscriber to purchase his or her DSL and ISP service from Vendor A, while contracting with Vendor B for all of the vertical services traditionally associated with a telephone call (e.g., callerID, call forwarding, three-way calling, voice mail, or advanced intelligent network services). This will have important consequences for the profitability of local service vendors. Currently, vertical services—such as custom calling features—are highly profitable, providing a subsidy to basic access (Huber, 1997). Increased competition for the provision of these vertical services should lead to a reduction in their profitability and a corresponding need to raise basic access prices closer to cost.

Competition in the provision of vertical services also raises technical questions. The ITU-T standards framework for VOIP, H.323, defines a gatekeeper function as the locus of vertical service provision. However, the standards as currently drafted do not easily accommodate the notion that subscribers might switch between multiple gatekeepers at will, or even use multiple gatekeepers simultaneously for different services. The use of multiple gatekeepers for different purposes would also complicate the provisioning of E.164 to IP directory services.

E.164 NUMBER MANAGEMENT

The average residence today has 1.2 telephone lines. However, in a VOIP environment, the number of distinct simultaneous voice calls is limited only by the upstream bit rate, which can easily support a half-dozen calls. Under these circumstances, residences may choose to acquire multiple E.164 numbers, one for each person or extension in the home. This could easily lead to even more rapid exhaustion of number space in the North American numbering plan. Alternatively, as with ISDN, a single E.164 phone number may be associated with multiple call appearances on different handsets. This conserves numbers at the expense of requiring consumers to learn some new conventions for call management.

AVAILABILITY OF DSL

The future of DSL as a service offering of the LECs faces significant regulatory uncertainty. The FCC has recently issued a notice of proposed rulemaking (NPRM) that presents the LECs with an uncomfortable choice: They can provision DSL service on an integrated basis, in which case they will be subject to the unbundling requirements which mandate that the service be made available to competitors at regulated total element long run incremental cost (TELRIC) rates. Or, they can provide DSL on an unregulated basis through a separate subsidiary. In that case, however, the separate subsidiary must acquire copper loops pursuant to the parent's unbundled network element (UNE) tariffs. This would force the LECs to become serious about UNE provisioning, which would then open the door to CLECs who wish to provide DSL service on a competitive basis using the same UNE loops (FCC, 1998).

DSL technology suffers from distance limitations. DSL is generally limited to loops shorter than 7 km without loading coils. Thus loops with loading coils, excessive bridge taps, or extended lengths must be rehabilitated before DSL can be deployed. Maximum achievable bit rates using DSL decline with distance. Furthermore, 25% of loops in the United States are served by DLC systems. These systems must be upgraded to support DSL line cards; and T1-over-copper feeder systems must be replaced by fiber before DSL can be offered to customers served by DLC.

CONCLUSIONS

Even without the additional revenue that a LEC might be able to earn for provisioning superior Internet service via DSL, in a greenfield setting, an integrated IP/ADSL network is cost-effective for small businesses simply on the basis of lower loop and interoffice transport costs. If the costs of premises gateways and DSLAMs continue to decline, we can expect that an IP/ADSL architecture will eventually be cost effective for residences as well.

Moreover, we have not examined any potential benefit to consumers of an IP/ADSL service in the form of lower interexchange calling costs. Elsewhere, we have shown that these can be substantial (Wanichkorn & Sirbu, 1998).

These results suggest that small businesses, Centrex users, and multiple-dwelling units might be the first customers targeted by an LEC seeking to migrate to an integrated packet network.

Finally, whereas the incumbent LECs are saddled with extensive investments in legacy circuit switching, a CLEC, unencumbered by past investments, can more readily choose to purchase only legacy loops as unbundled network elements while investing in new-generation IP-based switching and transport systems. RG and DSLAM products designed for CLEC use in this manner are already beginning to appear in the marketplace (Biagi, 1998).

ACKNOWLEDGMENTS

Support for this research was provided in part by a grant from Bellcore. The opinions expressed in this chapter are those of the authors and do not reflect the views of Bellcore or Carnegie Mellon.

REFERENCES

ADSL Forum. (1997). *ADSL forum system reference model* (Tech. Rep. No. TR–001). Mountain View, CA.

Biagi, S. (1998, December 7). Jetstream fills out DSL line. *Telephony*, pp. 14ff.

Bill, H. (1998). Home network. *PC Magazine, 17*(13), p. 209.

Clark, D. (1998). A taxonomy of Internet telephony applications. In J. K. MacKie-Mason & D. Waterman (Eds.), *Telephony, the Internet, and the media: Selected papers from the 1997 Telecommunications Policy Research Conference* (pp. 157–176). Mahwah, NJ: Lawrence Erlbaum Associates.

FCC. (1998). *NPRM: Deployment of wireline services offering advanced telecommunications capability* (CC 98-147). Washington, DC.

Freed, L. (1998, September 8). Networks made easy. *ZDNET Small Business Advisor.* Available http://www.zdnet.com/smallbusiness/stories/general/0,5821,339250,00.html.

Fryxell, D. (1998, June). *Analysis of ATM/ADSL architectures for a public broadband network from an economic and public policy perspective.* Paper presented at the Internet Telephony Consortium Semiannual Meeting, Helsinki, Finland.

Hatfield Associates. (1997). Hatfield Model Release 3.1. Boulder, CO.

Hawley, G. (1997, March). System considerations for the use of xDSL technology for data access. *IEEE Communications Magazine, 35*(3), pp. 56-60.

Holliday, C.R. (1997). The residential gateway. *IEEE Spectrum, 34*(6), 29–31.

Huber, P. (1997, November 4). Local exchange competition under the 1996 Telecom Act: Red-lining the local residential customer. Available http://www.cais.com/huber/redline/files.htm.

Humphrey, M., & Freeman, J. (1997, January–February). How xDSL supports broadband services to the home. *IEEE Network, 11*(1), pp. 14–23.

Leida, B. (1998). *A cost model for Internet Service Providers: Implications for Internet telephony and yield management.* Unpublished master's thesis, Massachusetts Institute of Technology, Cambridge, MA.

Morgan, S. (1998, January). The Internet and the local telephone network: Conflicts and opportunities. *IEEE Communications Magazine, 36*(1), pp. 42-48.

Niccolai, J. (1998, June 29). Vendor groups to ease home networking. *Infoworld*, p. 68.

Ohr, S., & Boyd-Merritt, R .(1998, March 9). Wireless-network debate hits home. *EE Times*, pp. 1, 6.

Okubo, S., Dunstan, S., Morrison, G., Nilsson, M., Radha, H., Skran, D.L., & Thom, G. (1997). ITU-T standardization of audiovisual communication systems in ATM and LAN environments. *IEEE Journal on Selected Areas in Communications, 15*(6), pp. 965–982.

Schechter, P. B. (1998, October). *Using cost proxy models to analyze universal service funding options.* Paper presented at the 26th Annual TPRC, Alexandria, VA.

Thom, G. A. (1996). H.323: The multimedia communications standard for local area networks. *IEEE Communications Magazine, 34*(12), pp. 52–56.

Turner, J. S. (1986). Design of an integrated services packet network. *IEEE Journal on Selected Areas in Communications, SAC-4*(8), pp. 1373–1380.

Wanichkorn, K., & Sirbu, M. (1998). *The economics of premises Internet telephony* (CMU Working Paper). Pittsburgh, PA: Carnegie Mellon University.

16

Implications of Local Loop Technology for Future Industry Structure

David D. Clark
Massachusetts Institute of Technology

This chapter explores the impact of the Internet on the deployment of technology for advanced residential network access. The shape of the future is certainly not clear, but certain considerations provide a basis for conjecture: the high cost of new wireline facilities, the emerging ability to provide higher quality Internet service over the existing wireline facilities of the incumbent local exchange carriers (LECs) and cable providers, the rapidly changing nature of the Internet and its service requirements, and the open nature of the Internet's interfaces, which tends to inhibit vertical integration of the Internet and the higher level services provided over it.

One possible outcome, considering these factors, is a future in which there is only a limited degree of competition in the provision of residential Internet service, and the degree of actual consumer choice changes rapidly due to the changing nature of the Internet, as well as the investment decisions of the facilities operators. Over the next decade, choice in residential Internet access could become as much of an issue as choice in local telephone service is today. Research and innovation in alternative modes of residential Internet access might improve the future options for competition.

INTRODUCTION

The industry structure surrounding the local loop is changing fast. The continuing process of deregulation combines with the advent of new service offerings such as the Internet to provide a powerful push for evolution. In attempting to examine this

283

market, the factors that are easier to assess are those that surround the more mature telephone and television services. Attention is naturally directed there because of the level of current and past investment and the size and influence of the industry players. Nonetheless, it is important to look at the possible shape of the industry that might emerge around new services, particularly the Internet, as they relate to the deployment of advanced local loop facilities. Such a look must be very speculative, but it can provide a common framework for discussion, and perhaps a common understanding of the range of options within which the future will evolve. It is possible that within a decade, society will be as concerned with the industry structure behind the Internet—the nature of competition, the range of consumer choice, the rate and level of investment in support of innovation—as we are today with the telecommunications industry.

It is difficult to predict the future course of the Internet. The Internet is a créature of the computer industry, and it evolves rapidly, as do all parts of that industry. It evolves in response to the emergence of new computer-based services, and the services and the Internet drive each other. The rapid rate of innovation interacts with the need for new investment to sustain the advances, and this interaction creates a future that is difficult to predict.[1] However, there are specific factors that seem to constrain the future, especially when looking at the local loop.

The key issues facing the potential provider of advanced access to the residence are as follows. To install a new generation of access technology that reaches the residence and small business is expensive. This level of investment is not likely to happen many times in a given location in any technology cycle, especially for wireline services that imply a large, up-front investment. The business case justifying any such investment will be built on three major service offerings—telephony, television, and Internet. The former two are relatively well understood, but the rapid evolution of the Internet makes it harder to predict the nature of the higher level services to be delivered there. At the same time, predicting the future Internet-based services is critical, because these determine the importance and utility of advanced access facilities. One must ask whether in the life of any current or new access facility, the Internet will evolve to deliver new services such as television and telephone service, or just continue to provide the "traditional" Web and e-mail. There is thus a tension between the apparent drive for the Internet to evolve and the difficulty of justifying the necessary investment.

The goal of this chapter is to focus on possible shapes of the industry surrounding the local loop that provides network access to the residence and small business, looking at the Internet as a shaping factor. The intention is to provide a

---

1. For a general discussion of the uncertain but inevitable evolution of the nation's communications infrastructure, see the report by the Computer Science and Telecommunications Board (1996).

framework for debate and identify certain constraints on the future, acknowledging that this sort of discussion is very speculative.

BASELINE: THE "OLD" STRUCTURE

To provide a baseline for discussion, Fig. 16.1 presents a simplified and abstracted illustration of the past, showing the three major communication services that reached the home and the technologies that carried them. Telephone was carried over copper pairs, radio over metropolitan broadcast, and television over broadcast or coaxial cable. There are two important points about this structure. First, there is a direct linkage between the delivery technology and the service (e.g., the provider who installed and operated copper pairs knew that the service being provided was telephone). This clarity in defining the line of business allowed the provider to design the system to optimize the known service and made it somewhat straightforward to construct a business plan for investment in infrastructure. Second, and related, the service and the technology were provided by the same company. The facilities provider and the service provider were vertically integrated.

FIGURE 16.1
Past Industry Structure of Residential Access.

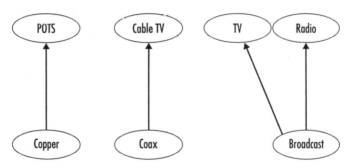

THE EMERGING STRUCTURE

Starting some time ago, this picture began to change. Divestiture, and the resulting recognition that it was potentially beneficial to be in multiple lines of business, caused the simple picture to evolve toward a more complex structure. Figure 16.2 represents a view of what we might expect in the near future, where solid lines represent what is actually available today, and dotted lines represent reasonable possibilities in the not too distant future. The services represented are the same with the addition of the Internet, and two new wireless delivery technology modes have been added, satellite and cellular (in contrast to single-tower metropolitan broadcast).

There are several points about Fig. 16.2. The first is that the strong vertical pairing of the first figure is replaced by a matrix structure. Many services are coming over several delivery technologies. Second, the Internet is in an interesting intermediate position. It is delivered over lots of technologies, and lots of services are (or can be) delivered over it.

FIGURE 16.2
Emerging Industry Structure of Residential Access.

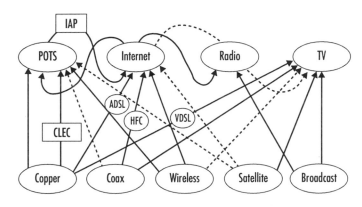

THE TELEPHONE INDUSTRY AS AN EXAMPLE

Figure 16.3 extracts from Fig. 16.2 the subset of links that relate to telephone service. The way the copper pair is being used has expanded, in that (at least in the United States) unbundling has been mandated as a way to increase competition. However, telephone service is also available using cellular communication, is now becoming available from satellite, has been provided in certain areas over coaxial cable, and is emerging as a service over the Internet. Although some of these modes are not yet technically mature (such as certain forms of Internet-based telephony[2]), and some like telephone over cable are being pushed by only some of the potential providers (for reasons that may have to do with economics and regulation as much as technology), this picture illustrates the complexity of the situation that the consumer, the industrial players, and the regulator must come to deal with.

---

2. Clark (1998) provided an analysis of different sorts of Internet telephony and the factors that limit the deployment of each.

FIGURE 16.3
Alternatives for Access to Telephone Service.

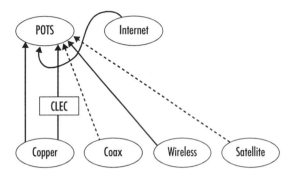

The regulatory situation is certainly more complex, in that it is now much less clear what (if anything) is to be regulated. In the old structure, the vertically integrated industries were easy to identify. However, in this picture, should one look at diversity in facilities, in higher level services, or some other criteria to assess the potential need for regulation? From the perspective of the consumer, the concern is quality and choice in the high-level services—telephone, television, and so on. The consumer is not directly concerned with the range of technology choices— how many fibers, coax lines, and copper pairs reach the house. This suggests that regulators should look at the higher level services to determine if consumer needs are being met by the competitive marketplace. However, the regulatory history, at least in the United States, applies a different regime to different providers based on the facilities they own. When a cable company and a telephone company propose to offer Internet service, they are subjected to different constraints because they are covered by different parts of the law.[3]

THE UNIQUE NATURE OF THE INTERNET

In Fig. 16.2, the Internet occupies a unique position. It can be provided over almost the full range of current and emerging local access technologies, and it can (or will in the future) be able to provide all of the enumerated services. It thus has the potential to be a universal means of facilitating the delivery of high-level ser-

---

3: A recent working paper from the Federal Communications Commission (FCC) Office of Plans and Policy (Esbin, 1998) offers a good discussion of the history and current status of regulation in this context, and uses the phrase *parallel universes* to describe the possible outcome of the straightforward application of today's regulation.

vices to the consumer. Figure 16.4 extracts from Fig. 16.2 the relationships relevant to the Internet.

This way of looking at the Internet is not new. A report from the Computer Science and Telecommunications Board (1994) illustrated the Internet as an hourglass, providing a single common means (by way of its standardized interfaces) to make a wide variety of technologies available to a wide variety of services. Tennenhouse, Lampson, Gillett, and Klein (1996) proposed the concept of a virtual infrastructure, where a range of technologies support a range of services via a single intermediate layer that they refer to as the *brokerage*. As Fig. 16.4 makes clear, this role as a cross-connect between technology and service is a very significant one for the Internet in the context of the local loop.

FIGURE 16.4
Potential Industry Structure Surrounding the Internet.

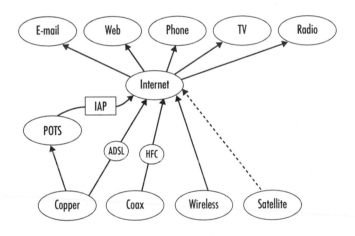

SERVICES OVER THE INTERNET

Looking first at the upper half of Fig. 16.4, how will higher level services over the Internet evolve, and what will be the implications? More specifically, what will inhibit or enhance the introduction and offering of each sort of service over the Internet? Second, are there forces that would favor or inhibit the integration of the higher level service with the provision of the basic Internet service itself?

This second question of who provides the higher level services and applications over the Internet is critical in predicting the future of the Internet and the industries that drive and shape it. The services that are illustrated in Fig. 16.2 do not include the "old" services that we traditionally associate with the Internet—e-mail, the World Wide Web, file transfer, and so on. For completeness, some of these are added to Fig. 16.4. To an experienced user of the Internet, these services are what the Internet is "for" today. I did not include them in Fig. 16.2 for a simple reason. These are not services that are today provided by a service provider. They do not represent services that somebody sells and consumers purchase. They come into existence through the combined efforts of all the producers and consumers of content and information exchange. People may make money by selling a particular file, but no service provider makes money by selling the file transfer service.

The new services that I illustrate have more direct analogs in the pre-existing world of consumer communication—telephone, radio and music delivery, and television. The cable television industry, if not the broadcast industry, sells access to television as a subscription service, which they present as a bundle with options, all of which they package and select for marketing.

An important speculation about the Internet is whether this more integrated model will emerge for some higher level Internet services, in which these services are bundled with the lower level delivery service for the Internet itself. Past experience would suggest that the answer is no. As noted earlier, current high-level services are not sold by Internet service providers (ISPs) as bundled products. Computer-to-computer Internet telephony is emerging as a collection of stand-alone software packages and network-based products being sold by independent third parties, not as a service being provided by ISPs (or anyone else). Other high-level services that are now evolving seem to have a similar structure.[4]

Internet Radio and Music

Internet radio, which is now emerging as a significant offering on the Internet, provides a chance to observe the development of an industry in its early stages. There are content packagers, such as Broadcast.com,[5] that are assembling a large amount of material and provisioning their own wide area infrastructure to better deliver it. At the same time, individual performers, radio stations, and other

---

4. The question of whether the Internet will lead to layered or integrated industry structure is discussed in a number of papers. Tennenhouse et al. (1996) argued that a horizontally layered system with decoupled layers will evolve naturally given the properties of digital technology, unless present convergence activities create a temporary monopoly condition. Gong and Srinagesh (1997) argued that the open structure, although a natural consequence of the open interfaces, may lead to reduced investment in facilities. Kavassalis, Lee, and Bailey (1998) discussed the factors that lead to different market structures and concluded that a layer such as the Internet can be sustained as an open interface, to some extent because Internet service, in contrast to raw capacity (e.g., fiber), is not a commodity but a differentiable product that will permit providers to set prices based on their value to the customers.

5. See http://www.audionet.com/about/ for information on this company and its offerings.

"small" sources of audio material are making their content available in piecemeal fashion. Whether the large or the small providers of audio content succeed in the market, the ISPs do not seem to be a significant provider of any of this service. Their only role is to upgrade the Internet to better carry this sort of content, and perhaps thus justify a higher fee for their Internet service.

Internet Television

This application of the Internet does not really exist at the time of this writing, but it is informative to speculate on the different forms it might take. A simple option is that television over the Internet works exactly the same way television does to-day—very high bandwidth access links are installed, and 50 to 100 or more channels are broadcast to the consumer over these links. However, an alternative model might be that the consumer subscribes to a number of sources of content, and these are downloaded in advance onto a local disk at the site of the consumer, where they can be watched at will. Some sorts of content, like the full-time news and weather channels, that provide highly repetitive material might achieve a tremendous reduction in required bandwidth to deliver their material by downloading the various pieces just once in the background, and then letting the viewer watch them later. This model of video distribution would contribute to much greater diversity of programming, because channels with only limited content (insufficient to fill a cable slot full time) could still develop a market.

This model of Internet television cannot instantly come into existence because it requires simultaneous evolution of the local loop, the consumer equipment (the successor to the set-top box), and the mode of content formulation and organization by the producer. This interaction illustrates the point that the Internet and the applications that run over it coevolve, which makes predicting the future (and especially its timing) very difficult. Depending on which model emerges, however, the communications technology that supports the Internet (e.g., satellites) might be subject to very different requirements—high-speed download of real-time video or "trickle charging" the consumer's equipment with prerequested content.

TECHNOLOGY BASE

Turning from the upper part of Fig. 16.4, which concerns higher level services, to the lower part, which concerns the delivery technology, there are again two questions to ask. First, to what extent are all of these technology options the same from the perspective of the user? Second, how rich will the competition be in providing them?

There is a wide range of delivery modes for Internet illustrated in Fig. 16.4. They differ in a number of respects. Some, like asymmetric digital subscriber lines (ADSL) and Internet protocol over cable, are capable of rather high-speed delivery, perhaps several Mb/s (at least "downstream" toward the consumer). Internet over dial-up modem, in contrast, is limited to no more than 56 Kb/s today and is not likely to get faster. Some forms of wireless service will be even slower than this. Different delivery modes differ not just in speed. For example, Internet over dial-up modem is only connected when the consumer makes a phone call for the purpose. Internet over cable and ADSL (high-speed Internet over copper) are services designed to be available at all times.

Do these differences matter? The answer is that it depends on the higher level service being used by the consumer. For e-mail, there is little compelling difference between a 56 Kb/s modem and a faster link. For cruising the Web, the increased speed seems to be valuable, and for Internet television, when and if that becomes significant, the 56 Kb/s modem will simply not be enough. Some forms of Internet telephony, in which calls are placed to the recipient directly over the Internet, are difficult to bring to market if the recipient is not connected to the Internet at all times. If the user must dial up in advance to receive a call, this prevents receiving a call without prearrangement. This limited service is hardly a replacement for traditional telephone service. On the other hand, using an "always on" Internet service such as is provided by Internet over cable, it is possible to receive an unanticipated call, the way the telephone system works today. So the features that the Internet customer will demand will depend on the higher level services that are currently popular. And if a majority of the users do not have a suitable Internet service, this can cause certain high-level services to stall in the market.

Competition in Providing Internet Access Service

Just as we are concerned today with competition in the provision of telephone service, it is important to inquire now as to what degree of competition might finally emerge in the provision of Internet service. There is no certain answer today, of course, but we can see the relationship between decisions now being made and the eventual outcome.

One fact that seems quite certain is that installing a whole new wireline facility is very expensive. It is not likely that there will be many new wires (or fibers) installed to the residence in any technology cycle. At the same time, there is anecdotal evidence that the higher speed Internet options such as Internet over cable (or more specifically cable modems over hybrid fiber-coax [HFC]) are proving sufficiently popular that they may come to represent a distinct variant of Internet access for which the slower options like dial-up do not provide a direct substitute. Today in the United States, there is typically one provider offering copper pairs for ADSL (the incumbent LEC) and one provider of cable in any given area. (The situation in other countries will vary, as different patterns of deployment and cross-ownership apply in different parts of the world.) These are the only two high-speed wireline infrastructure options currently in the picture. So one extreme for future

Internet service is that high-speed Internet service is provided by a duopoly consisting of the current LEC and the current cable provider. Unless there is some business or regulatory pressure to move away from this outcome, it is a likely one.

There are other outcomes that are not so extreme. The LEC might sell an ADSL service and permit the consumer to select from a number of competing Internet services over that ADSL link. Or the LEC might be forced to unbundle the copper loop for ADSL service, by analogy with the current approach to service competition for telephone service.[6]

Although it is not possible to predict with certainty how these options might evolve, one can look at the current approach of the cable industry for a first hint. Currently, those cable providers who choose to sell Internet service over their cable do so by offering the consumer a bundled Internet service option, which they provide and sell as a part of their overall service product. There has been no tendency to give the consumer a choice of ISPs over their cable infrastructure. Were the LECs to follow this model, a duopoly in high-speed Internet service would be the outcome.

If high-bandwidth applications of the Internet become popular, so that the dial-up service becomes a second-tier service for customers interested in low price rather than service quality, the current very competitive market for consumer Internet access over modem will become squeezed into one low-value corner of the market, with the high end concentrated in the facilities-based providers, of which there might only be two. This sequence of events would signal a major transformation of the consumer Internet service industry.

The technical innovation most likely to alter this picture would be the emergence of some wireless service with enough bandwidth to compete with the performance of the wireline solutions. But this sort of service raises serious technical challenges, including the availability of sufficient suitable wireless spectrum, the difficulty of achieving the requisite bandwidth to the user, the need to provide the "always on" form of the service, and so on. It may be that if the duopoly as the final outcome is not considered an adequate range of choices for the consumer, some intervention in the market may be required.

Hybrid Technology

One interesting issue that is only now emerging is that it may be possible to construct a superior Internet service by using more than one sort of residential access technology at a time. For example, one variant of Internet access is provided today

---

6. The recent filing by the NTIA (1998) to the FCC advances both these options as desirable outcomes and supports regulatory unbundling of DSL facilities. This suggests that they believe that regulation is necessary, even at this early stage of the emerging market for advanced services, to mitigate the power of the facilities owner.

using satellite or cable in one direction and telephone links in the other. In the future, we may see more novel hybrids, for example involving low- and high-orbit satellites. Because, traditionally, one industrial player has installed one sort of technology, these hybrid options will force some sort of relationship between multiple players to provide the overall service that the consumer purchases.

A NEXT-GENERATION LOCAL LOOP TECHNOLOGY?

There is continued speculation that some form of advanced access technology might be widely installed to the residence; for example, fiber to the curb or fiber to the home. At present, there does not seem to be any widespread planning or investment in these next-generation technologies. There are a number of observations that can be offered concerning this situation. One is that the expense is such that the typical consumer will not see a high level of competition in this offering. It is quite possible that there would be at most one version of a next-generation wireline service for most consumers. Second, any such investment would almost certainly be motivated by the desire to get into as many high-level businesses as possible—telephone, television, Internet, and so on. So were this deployment to happen, it would represent a rather complicated business situation. On the one hand, it might serve to increase the competition in all of these higher level services. On the other hand, it might represent a noncompetitive presence in the access market that might drive the less capable technologies from the market and leave the consumer with insufficient choice in the basic access service. That outcome might lead to regulation of the new access facilities, specifically enforced unbundling of the new facilities so that competitive providers of telephone, television, and Internet are assured access. Certainly, anticipation of this regulatory outcome would have a chilling effect on the business plans of potential investors.

Since more than 40% of U.S. homes have personal computers, it is a plausible guess (but still just a guess) that at the right price there would be similar demand for high-bandwidth advanced network access, even if all it did was improve the utility of the computer by enabling a better version of Internet service. In fact, if a whole neighborhood is wired at once, the cost to each household might be the same magnitude as the purchase of a personal computer. However, individual consumers cannot make independent decisions to have advanced wireline facilities installed. To keep the cost of installation at a reasonable level, it is necessary to wire (or rewire) a whole neighborhood at the same time. Thus, collective rather than individual decision making is necessary.

Given the risks to the private investor and the inability of the individual consumer to act independently, it is possible that the future picture is one in which the access technology is a recognized monopoly or a nonprofit or government-sponsored facility, but there is open competition for all the services that run on top of it, including telephone, television, and the Internet. Currently, there are a number of local municipal governments experimenting with the installation of advanced access facilities, such as fiber to the home.[7] Although the approaches vary widely in design, including both the services offered and the model of financing, many require the consumer to pay a significant up-front cost. By asking the consumer to bear some of the up-front cost, the financial risk to the installer is reduced. At the same time, the nonprofit or governmental player makes possible the necessary collective action so that whole neighborhoods or communities can be upgraded at once.

Regions of different demographics, regulatory history, and physical conditions offer different opportunities for competition and can support different technical options. The northeastern part of the United States, which mostly has a dense tree cover, has fewer options for wireless deployment than parts of the West, because the water in the tree leaves is opaque to many of the frequencies used for broadband wireless access. Multifamily dwellings can have a lower cost to wire than the dwellings on the fringes of the suburbs, and thus might better sustain competition in access options. Any speculations about the future, whether business plans or options for regulation, must take into account that different conditions may prevail at different times and places. We are not likely to see either a uniform monopoly or successful universal competition in advanced services. Specific providers may find themselves in different states of competition in different parts of their operating range. These realities will raise new issues for regulators and policy planners.

CONCLUSIONS

Wholesale installation of wireline access technology to the residence is expensive enough that we cannot expect a rush to enter this market. At the present time, there are two incumbents, the LEC and the cable provider. Both are moving to enter new service markets, in particular the Internet market. A number of factors will shape

---

7. Examples include Ashland, Oregon (see http://www.projecta.com/afn/), Palo Alto, California (see http://www.city.palo-alto.ca.us/palo/city/utilities/fth/index.html), and Glasgow, Kentucky (see http://www.glasgow-ky.com/). The term *community networking* is used to cover a range of activities from municipal wiring to library-based access and community Web pages. Useful sites that relate to community networking include a website maintained by David Pearah at the Internet & Telecoms Convergence Consortium at MIT (see http://itel.mit.edu/communitynetworks/links.html), the site of the Center for Civic Networking (see http://civic.net/lgnet/telecom.html), the Community Networking page of Big Sky Telegraph (see http://macsky.bigsky.dillon.mt.us/community.html), a resource page from the Association of Bay Area Governments (see http://www.abag.ca.gov/bayarea/telco/other.html), the Directory of Public Access Networks maintained by the Council on Library and Information Resources (see http://www.clir.org/pand/pand.html), and an online guide maintained by Paul Baker at George Mason University (see http://ralph.gmu.edu/~pbaker/index.html).

the future of the local loop. In the short run, there do not seem to be any serious plans to install additional wires (or fibers) to the home.

One possible outcome is that there are two providers of high-bandwidth Internet service, the incumbent LEC and cable provider. Although there will be other forms of Internet access (wireless, satellite, etc.) these may be sufficiently different in features such as bandwidth and continuous availability that they do not directly substitute for the high-bandwidth wireline solutions. The result is a duopoly in the provision of residential Internet access.

If this outcome is considered undesirable, one way to mitigate it (other than regulatory intervention) would be to encourage research in alternative delivery technologies, including high-bandwidth wireless, and hybrid models that use more than one technology to build a single, high-performance Internet service. However, exactly which forms of Internet service are in practice substitutable will depend on which higher level applications become popular, and that popularity can and will change over time. It is thus plausible to anticipate that the competitive breadth of residential access to the Internet may change with the pace of the evolution of higher level services, which can happen much faster than the pace of infrastructure investment.

The implication of the Internet for consumer access to higher level services is that there may be increased competition in the provision of these services, including those such as telephone and television that are limited in competitive breadth today. This derives from the open character of the Internet design that militates against vertical integration of the ISP and the higher level service provider.

ACKNOWLEDGMENTS

This work was supported by the MIT Internet & Telecoms Convergence Consortium and its industrial sponsors.

REFERENCES

Clark, D. (1998). A taxonomy of Internet telephony applications. In J. K. MacKie-Mason & D. Waterman (Eds.), *Telephony, the Internet and the media: Selected papers from the 1997 Telecommunications Policy Research Conference* (pp. 157–176). Mahwah, NJ: Lawrence Erlbaum Associates.

Computer Science and Telecommunications Board. (1994). *Realizing the information future: The Internet and beyond.* Washington, DC: National Academy Press.

Computer Science and Telecommunications Board. (1996). *The unpredictable certainty: Information infrastructure through 2000.* Washington, DC: National Academy Press.

Esbin, B. (1998). *Internet over Cable: Defining the future in terms of the past* (Working Paper No. 30). Washington, DC: FCC Office of Plans and Policy.

Gong, J., & Srinagesh, P. (1997). The economics of layered networks. In L. McKnight & J. Bailey (Eds.), *Internet economics* (pp. 63–75). Cambridge, MA: MIT Press.

Kavassalis, P., Lee, T. Y., & Bailey, J. P. (1998). Sustaining a vertically disintegrated network through a bearer service market. In E. Bohlin & S. L. Levin (Eds.), *Telecommunications transformation: Technology, strategy, and policy* (pp. 151–172). Washington, DC: IOS Press.

NTIA. (1998). Filing before FCC by NTIA concerning Section 706 of the Telecommunications Act of 1996. Available http://www.ntia.doc.gov/ntiahome/fccfilings/sec706.htm

Tennenhouse, D., Lampson, B., Gillett, S., & Klein, S. (1996). Virtual infrastructure: Putting information infrastructure on the technology curve. *Computer Networks and ISDN Systems, 28,* 1769–1790.

17

The Internet and "Telecommunications Services," Access Charges, Universal Service Mechanisms, and Other Flotsam of the Regulatory System

Jonathan Weinberg
Wayne State University

As digitization and packet switching revolutionize communications networks, U.S. communications law faces a dilemma: How can it reconcile an old law that contains distinct regulatory structures for telephony, broadcasting, cable television, and satellites, and that leaves Internet protocol (IP) transmission largely unregulated, with a future in which voice, video, text, and data will simply be different forms of information to be transmitted using IP or other packet-switched protocols? In particular, to what extent should (or can) regulators impose existing telecommunications regulation, including the snarl of cross-subsidies that dominates the telephone system, on IP networks?

At the heart of existing telecommunications regulation is the distinction between *telecommunications service,* whose providers are subject to Federal Communications Commission (FCC) common-carrier regulation and must pay a percentage of revenues to a federal universal service fund, and *information services,* whose providers are subject to no such obligations. The dividing line Congress and the FCC have drawn between these categories, though, is no longer workable. We should redraw that line in the modern communications context, so as to focus regulatory obligations on the underlying transport. The FCC must approach existing subsidies to make its regulation minimally distorting and technology neutral.

INTRODUCTION

Packet-switched networks are taking over, and the communications world is changing. The traditional communications world relies on distinct infrastructures for each communications service. Voice travels over a nationwide wired, intelligent, circuit-switched network, with a single 64 Kbps voice channel set aside for each call. Video moves over a separate system of terrestrial broadcast stations, supplemented by coaxial cable or hybrid fiber-coax networks carrying video programming from a cable headend to all homes in a given area. Data is piggy-backed onto the voice network via an awkward kluge, under which the information is converted from digital to analog form and back again.

Digitization and packet switching, though, have the potential to change that traditional design. One can convert the information transmitted via any communications service—whether it be voice, video, text, or data—into digital form. Packet switching, with or without the use of IP, enables the transmission of that digitized information across different networks without regard to the underlying network technology. This means that the digitized information corresponding to any service can be transmitted over any physical infrastructure—copper wires, fiber, hybrid fiber-coax, microwave, or direct broadcast satellite. Proprietors of copper (or hybrid fiber-coax, or wireless) infrastructure can offer services not previously associated with those physical facilities, and new services can be delivered, via the Internet, over any physical facilities supporting high-speed data transmission.

Both local exchange carriers (LECs) and cable operators are now entering the market to provide high-speed data services. Consumers with Internet access can engage in real-time voice transmission via IP. Cable operators are exploring the provision of voice telephony, via IP, over cable facilities. New services, including video, can be offered over various facilities; all that is necessary is bandwidth. Increasingly, firms are designing nationwide, packet-switched, backbone networks to carry that traffic. The networks are not designed to support any particular service; they carry whatever information is necessary for the service the consumer wants.

These developments, however, give rise to a regulatory dilemma. U.S. communications law has developed along service-specific lines. It has formulated complex and distinct regulatory structures covering telephony (wired and wireless), broadcasting, cable television, and satellites. It has so far left IP transmission largely unregulated. As those technologies become no longer separate, we need to figure out what to do with the old regulatory structures.

In this chapter, I focus on one aspect of that problem: To what extent should (or can) we impose legacy telephone regulation on IP networks? As a broad-brush matter, it seems plain that it would be a Very Bad Idea either (1) to impose such regulation; or (2) not to impose it. Imposing legacy regulation on IP networks seems like a bad idea because that regulation was not designed for packet-switched networks. It was developed to fit a circuit-switched world, served mostly by monopoly telephone service providers, and it is characterized by extensive cross-subsidies and a general disregard of innovation and competitive markets. Not imposing regulation, though, seems untenable as well: IP and conventional networks are merging. To maintain extensive regulation of the circuit-switched world and minimal regulation of the IP world will simply invite arbitrage and undercut the legitimate policy goals of the old system.

This problem is made more difficult by the snarl of cross-subsidies that make up much of modern telephone regulation. Telephone pricing today is characterized by a tangle of subsidies: some federal, some state; some explicit, some implicit.[1] On the federal level, the government administers explicit subsidies through universal service contributions and disbursements.[2] The largest explicit federal subsidy is the "high-cost" component of the Universal Service Fund. The high-cost fund subsidizes telephone companies in rural and other high-cost areas, where local loop costs are so high that many users would drop off the network rather than shoulder those costs themselves. It is funded through "contribut[ions]" (as the statute puts it) by interstate telecommunications providers;[3] in the fourth quarter of 1998, those firms will pay about 2.5% of their interstate and international end-user revenues for that purpose. The federal government, though, also implements implicit subsidies through the interstate access charge system,[4] and through geographic averaging of interstate long-distance rates.[5]

States typically administer a maze of implicit subsidies via geographic averaging of local telephone rates, business-to-residential subsidies, and the pricing of vertical features, intrastate access, and intrastate toll. The most important such device

---

1. *Implicit subsidy,* in this context, means that "a single company is expected to obtain revenues from sources at levels above 'cost'…and to price other services allegedly below cost." Federal-State Joint Board on Universal Service, Report and Order, 12 FCC Rcd 8776, 8784 n. 15 (1997) (*Universal Service Order), appeal pending sub nom.,* Texas Office of Public Utility Counsel v. FCC, No. 97-60421 (5th Cir.).

2. The federal government also oversees other universal service programs: the Lifeline and Linkup programs, targeted toward low-income consumers, and a program designed to connect schools, libraries, and rural health care providers to the Internet. For the fourth quarter of 1998, the Universal Service Administrative Company estimated $422,500,000 demand (and $423,700,000 total costs) for the high-cost program, amounting to 48% of all Universal Service Fund expenses. The schools and libraries fund is projected to make up 37%; the Lifeline and Linkup programs, 13%; and rural health care, 3%. See Proposed Fourth Quarter 1998 Universal Service Contribution Factors Announced, CC Docket No. 96-45 (rel. Aug. 18, 1998). Available http://www.fcc.gov/Bureaus/Common_Carrier/Public_Notices/1998/da981649.html.

3. 47 U.S.C. § 254(d).

4. See 47 C.F.R. Part 69.

5. See 47 U.S.C. § 254(g).

for the states is geographic rate averaging: High-density urban areas, where costs are lower, underwrite the provision of service to low-density, high-cost rural areas.[6]

The FCC, thus, must face these questions: To what extent should the Internet, and IP networks generally, be brought into the web of subsidies that characterize much of modern telephone regulation? What are the consequences if they are not?

In this chapter, I focus my attention on explicit federal universal service subsidies and, to a lesser extent, on the interstate access charge system. I suggest that the distinction between telecommunications and information service embedded in current law is not appropriate for defining universal service obligations. Tying universal service obligations to the ownership of network facilities is more appealing in several respects, but there are significant questions left to resolve. I urge that Internet service providers (ISPs) should not pay interstate access charges: We should not bring IP networks into implicit subsidy arrangements that we are trying to eliminate in the circuit-switched world. Nonetheless, the FCC should consider announcing an obligation that ISPs pay a federally tariffed charge for connection to the local network conditioned on the existence of meaningful competition in packet-switched transport in that geographic market.

EXPLICIT SUBSIDIES

The April 1998 Report to Congress on Universal Service

The FCC squarely confronted the relationship of IP-based service providers with the federal universal service system in 1998.[7] An appropriations rider had directed the agency to undertake a detailed review of its definitions of the terms *information service, telecommunications,* and *telecommunications service* (among others) in the Telecommunications Act of 1996; the application of those definitions to "mixed or hybrid services" (referring in part to Internet access services and IP telephony); and "the impact of such application on universal service definitions and support."[8]

The Congressional directive focused on the definitions of telecommunications and information service because those definitions trigger regulatory obligations. Providers of telecommunications on a common-carrier basis are required to pay a percentage of their end-user revenues as a contribution to the Universal Service Fund[9] and are subject to extensive obligations under Title II of the Communica-

---

6. See *Universal Service Order,* 12 FCC Rcd at 8784.

7. See generally Federal–State Joint Board on Universal Service, 13 FCC Rcd 11501 (1998) (*Report to Congress on Universal Service*).

8. Departments of Commerce, Justice, and State, the Judiciary, and Related Agencies Appropriations Act, 1998, Pub. L. No. 105–119, 111 Stat. 2440, 2521–2522, § 623.

9. See 47 U.S.C. § 254(d); *Universal Service Order,* 12 FCC Rcd at 9205–12; *supra* note.

tions Act.[10] Providers of telecommunications on other than a common-carrier basis are exempt from almost all Title II requirements, but may nonetheless be required to pay into the Universal Service Fund.[11] By contrast, the mere provision of an information service triggers no universal service contribution or other regulatory requirement.[12]

The Congressional directive stemmed from the view of some senators that all IP-based services should be deemed to involve telecommunications, and that providers of such services presumptively should make universal service contributions and be subject to the other requirements of Title II. Those senators expressed concern that the contrary view would endanger the Universal Service Fund, as telephone traffic shifts from conventional to IP networks.

The FCC dutifully set about the task of categorizing IP-based services as telecommunications or information services. The agency's first step was a finding that Congress, in referring to telecommunications and information services, had intended to build on the FCC's 1980 *Computer II* framework, which characterized all services involving the transmission of information as "basic" or "enhanced."[13] *Computer II* had defined *basic* transmission services as the offering of "pure transmission capability over a communications path that is virtually transparent in terms of its interaction with customer supplied information,"[14] while classifying any more elaborate offering over the telecommunications network as an *enhanced* service.[15] Enhanced service providers, the FCC stated, should not be subject to regulation under Title II.[16] That approach was successful in spurring innovation and competition in the enhanced services marketplace: Government was able to maintain its control of the underlying transport, sold primarily by regulated monopolies, while eschewing any control over the newfangled, competitive "enhancements."

---

10. 47 U.S.C. §§ 201–76.

11. *Id.* § 254(d).

12. See *Report to Congress on Universal Service*, 13 FCC Rcd at 11515–16.

13. *Id.* at 11520–26; *see* Amendment of Section 64.702, 77 FCC 2d 384 (1980) (*Computer II*), *on recon.*, 84 FCC 2d 50 (1980), 88 FCC 2d 512 (1981), *aff'd sub nom.* Computer and Communications Industry Ass'n v. FCC, 693 F.2d 198 (D.C. Cir. 1982), *cert. denied*, 461 U.S. 938 (1983).

14. *Id.* at 419–20.

15. *Id.* at 420; *see also* 47 C.F.R. § 64.702(a).

16. The agency reasoned that enhanced services involve "communications and data processing technologies...intertwined so thoroughly as to produce a form different from any explicitly recognized in the Communications Act of 1934," and that enhanced service providers were not "common carriers" within the meaning of the Act. *Computer II*, 77 FCC 2d at 430–32. The agency's rule of nonregulation, though, included one major caveat: Ma Bell and her descendants, when they sought to offer enhanced services, were subject to restrictions designed to ensure that they did not leverage their monopoly power.

The 1996 Telecommunications Act contains no references to basic and enhanced services; instead, it characterizes communications services as telecommunications or information service.[17] The Act defines telecommunications, though,
in a manner strongly reminiscent of the basic services category, as "the transmission, between or among points specified by the user, of information of the user's
choosing, without change in the form or content of the information as sent and received."[18] It defines information services in a manner reminiscent of enhanced
services, to include "the offering of a capability for generating, acquiring, storing,
transforming, processing, retrieving, utilizing and making available information
via telecommunications."[19]

In its 1998 Report to Congress on Universal Service, the FCC found that Congress in the 1996 Act intended telecommunications and information service to be
mutually exclusive categories roughly paralleling basic and enhanced services.[20]
The FCC went on to find that Internet access services were information services
(and that the providers of such services therefore were under no obligation to make
direct payments to the Universal Service Fund). By contrast, the FCC classed the
provision of pure transmission capacity to Internet access and backbone providers
as telecommunications.

The agency ran into some difficulty when it sought to classify IP telephony services—that is, services enabling real-time voice transmission using IP. The FCC
began by addressing "computer-to-computer" IP telephony, in which individuals
use software and hardware at their premises to place calls between two computers
connected to the public Internet. In that context, the FCC stated, it need not decide
whether there was telecommunications taking place; Title II requirements (including universal service payment obligations) would not apply in any event. Title II
obligations, the agency explained, apply only to the "provi[sion]" or "offering" of
telecommunications. When a user, with an ordinary Internet connection through
his or her ISP, uses Internet telephony software to enable real-time voice communication between the user's computer and that of a fellow enthusiast, the ISP may
not even know that the subscriber's packets are carrying voice communications.
The ISP is not, in any meaningful sense, "provid[ing]" the voice telephony to that
subscriber, and cannot be made subject to Title II on that basis.[21]

The agency was unable to be so definite, though, with respect to "phone-to-
phone" IP telephony services. Those are services in which a customer places a call,

---

17. Those categories originated in the 1982 Modification of Final Judgment (MFJ) ending the antitrust suit between the U.S. government and AT&T. See United States v. American Tel. & Tel. Co., 552 F. Supp. 131, 226-32 (D.D.C. 1982), *aff'd sub nom.* Maryland v.
United States, 460 U.S. 1001 (1983); *Report to Congress on Universal Service*, 13 FCC
Rcd at 11514, 11521–22.

18. 47 U.S.C. § 153(43).

19. *Id.* § 153(20).

20. *Report to Congress on Universal Service*, 13 FCC Rcd at 11511.

using an ordinary telephone and the public switched telephone network, to a gateway device that packetizes the voice signal and transmits it via IP to a second gateway, which reverses the processing and sends the signal back over the public switched network to be received by an ordinary telephone at the terminating end. The FCC was unable to reach a conclusion as to the proper classification of such services, stating only that "[t]he record currently before us suggests that certain forms of 'phone-to-phone' IP telephony lack the characteristics that would render them 'information services' within the meaning of the statute, and instead bear the characteristics of 'telecommunications services.'"[22] It deferred any "definitive pronouncements" on phone-to-phone IP telephony to an unspecified later proceeding.

The Report to Congress on Universal Service stressed the FCC's position that the growth of IP-based services would buttress universal service, not undercut it. Notwithstanding that Internet access providers are not required to make universal service payments, they are major users of telecommunications, and thus make "substantial indirect contributions to universal service" in the prices they pay to purchase telecommunications.[23]

The Breakdown of the Telecommunications–Information Service Distinction

It should not be surprising that the FCC had trouble with the characterization and regulatory obligations of IP telephony providers. On the one hand, the agency was surely correct that Title II obligations should not leap into existence simply because a consumer transmits voice rather than, say, graphics, over an IP connection. It would be highly problematic to treat packets differently just because they carried voice rather than some other sort of information. More important, as the FCC noted, in the typical "computer-to-computer" case, the customer is buying only Internet access, rather than an IP telephony service as such. In a wide variety of IP-based services, further, the customer will be receiving enhanced functionality that goes far beyond the plain-vanilla transmission that constitutes telecommunications.

On the other hand, it is also problematic if the provider of a service that looks and feels to the user just like conventional telephony is subject to regulation far different from that imposed on conventional telephony providers. In particular, it would be odd and unhelpful if huge regulatory distinctions should turn on the question of whether a vendor transports an intermediate leg of its telephone calls via IP or via some other packet-oriented technology. Conventional telecommunications carriers are increasingly using asynchronous transfer mode (ATM), a dif-

---

21. *Id.* at 11543. If the user is reaching his or her ISP over a dial-up telephone connection, then the telephone company is providing the user with telecommunications, but that service is wholly distinct from the IP telephony functionality. *Id.* at 11534 n.187.

22. *Id.* at 11508.

23. *Id.* at 11503–04.

ferent packet-oriented communications technology, in their networks, and it is completely accepted that the use of ATM to transmit a telephone call does not render the carrier an information service provider.

To accommodate both of these concerns, one must devise a way of distinguishing those forms of IP telephony that should be subject to regulation from those that should not; but that turns out to be troublesome indeed. The FCC in the Report to Congress on Universal Service suggested the possibility of subjecting to Title II regulation IP telephony services in which the provider

> 1) . . . holds itself out as providing voice telephony or facsimile transmission service; 2) ...does not require the customer to use CPE [customer premises equipment] different from that CPE necessary to place an ordinary touch-tone call (or facsimile transmission) over the public switched telephone network; 3)...allows the customer to call telephone numbers assigned in accordance with the North American Numbering Plan, and associated international agreements; and 4)...transmits customer information without net change in form or content.[24]

The effect of these requirements would be to regulate an IP telephony service as telecommunications if the customer's signal travels in unpacketized form over the public switched network to a gateway (as in conventional phone-to-phone service), but not if it is packetized in the CPE (as in computer-to-computer service).

It is doubtful that that would work very well. Consider a telephone handset that packetizes the customer's voice signal and sends the packets via IP to an Internet telephony service provider, but that nonetheless looks and acts, from the user's perspective, like a conventional telephone. If a service should rely on such equipment, it is not obvious what policy goals would be served by treating that service differently from phone-to-phone IP telephony as defined.[25]

The difficulty, further, extends beyond the particular definition suggested in the Report to Congress on Universal Service. What if the phone-to-phone IP telephony provider adds just a dab of functionality—say, it not only enables two people to talk, but automatically records the conversation and makes it available via streaming audio on a website? Or—to enable anybody to be a talk show host—it allows the originator to conduct a conference call with three or four people, while

---

24. *Id*. at 11543–44.

25. See *id*. at 11622–24 (dissenting statement of Commissioner Furchgott-Roth). Indeed, consider business telephone users served by switchboard or Centrex systems. Should the policymaker apply one regulatory paradigm if calls from the business's telephones are directed to an IP gateway on the public switched network, but another if the switchboard serving the business itself serves as such a gateway? Why?

allowing any member of the public to dial in and listen?[26] Looking to the 1996 Act definitions, it seems plain that the recording, storage, and rebroadcast of the conversation in the first example involves enhanced functionality and constitutes an information service; the more difficult question is whether we have one service or two. That is, does the example involve a single information service, or a plain-vanilla telephony service (telecommunications) combined with a separate transcription service (information service)? Similarly, the service in the second example does not appear to qualify as telecommunications, which the 1996 Act defines as "the transmission, between points specified by the user, of information of the user's choosing, without change in the form or content of the information as sent and received"—because the transmission does not seem to be "among points specified by the user."[27] Yet should we therefore characterize the overall service as an information service, or can one again find a regulated telecommunications service by dividing the offering into two?

The anomaly here is that it is easy to build functionality into IP-based services, yet under the telecommunications–information service dichotomy, an IP-based service will be deemed telecommunications and thus will be subject to regulation and universal service obligations, only if it sufficiently limits what the user can do. If the same service gets a software upgrade and allows the user to do more elaborate things, it becomes an information service and escapes regulation—unless the new component is deemed a separate service. However, we do not have any rules for deciding when that should be so. It is hard to see why any of this makes sense.

IP telephony, further, presents another puzzle: Under the FCC's current definitions, phone-to-computer and computer-to-phone IP telephony both appear to be information services. In each case, the gateway is providing protocol conversion and processing (translating from unprocessed voice to a series of IP packets, or vice versa); under established rules, that enhanced functionality pulls the service out of the realm of simple telecommunications.[28] Yet put those two services together, and what do you have? Any protocol conversion taking place at one point in the call is undone at another; established law suggests that the concatenated services are mere telecommunications.[29] It is hard to know what to do with that, since the firms providing the two services may not, in a distributed environment, even be aware of one another.[30] Bottom line: The telecommunications–information ser-

26. These examples are Mike Nelson's.

27. 47 U.S.C. § 153(43).

28. See Implementation of the Non-Accounting Safeguards of Sections 271 and 272, Report and Order, 11 FCC Rcd 21905, 21955–58 (1996) (*Non-Accounting Safeguards Order*) (in general, services involving protocol processing fall within 47 U.S.C. § 153(20)'s definition of "information service," because they offer "a capability for…transforming [and] processing…information via telecommunications"), *on recon.*…, 12 FCC Rcd 2297, *aff'd sub nom.* Bell Atlantic Tel. Cos. v. FCC, 131 F.3d 1044 (D.C. Cir. 1977).

vice boundary does not seem to divide up the world of IP-based services in any especially useful way.

Nor are these problems limited to IP telephony. Consider a rather more important finding of the Report to Congress on Universal Service: that Internet access is an information service. That conclusion seems vulnerable, outside of the dial-up context. One of the defining characteristics of IP is that an IP network itself displays no intelligence; it only passes information transparently from one edge to another. In a phrase, the network provides only *commodity connectivity.*[31] All of the intelligence and enhanced functionality—the storage and manipulation of user information—takes place at the edges of the networks (i.e., either before or after the information is transmitted from origin to destination). Simple IP transmission, thus, seems like a classic example of "transmission, between or among points specified by the user, of information of the user's choice, without change in the form or content;" that is, telecommunications. Indeed, the FCC has said essentially that about other packet-based services.[32]

The Report to Congress on Universal Service bases its finding that Internet access is an information service largely on the fact that ISPs run mail servers, host Web pages, offer Usenet news feeds, operate caches, and engage in other computer-mediated activities that go beyond simple transmission of packets. However, not all customers require these services. Where the customer is a corporate intranet, it will maintain its own mail and Web servers. The Internet access provider likely will provide nothing except pure transmission and routing of packets within its internal network and connection to the larger Internet. In such a case, it seems hard to avoid the conclusion that the ISP is offering telecommunications: It is providing transport and nothing else.

The notion that pure Internet connectivity therefore is telecommunications within the meaning of the 1996 Act, though, is troubling on a fundamental level. It expands the scope of services subject to Universal Service Fund obligations without any policy-oriented understanding of why that should be necessary or desirable. Put another way, it extends old rules to the Internet without adequate consideration of whether that is a good thing.

------

29. See Deployment of Wireline Services Offering Advanced Telecommunications Capacity, CC Docket No. 98–147 (Aug. 7, 1998), at ¶ 35 n. 57 (*Section 706 Order and NPRM*); *Report to Congress on Universal Service*, at 11526 & n. 106; *Non-Accounting Safeguards Order*, 11 FCC Rcd at 21958.

30. I am indebted to Stagg Newman for his emphasis of this point.

31. The phrase is from David S. Isenberg, Dawn of the Stupid Network, available http://www.isen.com/papers/Dawnstupid.html.

32. See Independent Data Communications Mfrs. Ass'n, 10 FCC Rcd 13717 (1995); cases cited in *Section 706 Order and NPRM* at ¶ 35 n. 56 and accompanying text.

Why the Telecommunications–Information Service Distinction Does Not Work

To understand why the telecommunications–information service distinction does not work in the IP context, it is useful to look back to *Computer II*. The *Computer II* categories, like their 1996 Act cognates, focused on service offerings. That is, the things being categorized were services, rather than (say) equipment, or capabilities. That is a natural way to divide up the world from a conventional telephony perspective; folks from the world of computer-to-computer communications, though, tend to use a different set of categories.

The computing world, in thinking about the communications process, tends to rely on the open systems interconnection (OSI) model, which organizes that process into layers.[33] The *physical* layer is concerned with the physical infrastructure over which the information travels; immediately above that are the *data link* layer, concerned with the procedures for transmitting data using a particular technology, and the *network* layer, concerned with the transfer of data between computers and routing. The *transport* layer defines the rules for information exchange and manages the reliable end-to-end delivery of information. The *session, presentation,* and *applications* layers focus on user applications; in particular, the applications layer contains the functionality for specific services.

The service offerings contemplated by *Computer II* cut across the layers of the OSI model. For example, the paradigmatic example of basic service (or telecommunications) is plain old telephone service (POTS), designed to enable ordinary voice communication. POTS constitutes a vertically integrated intertwining of components from various layers. It relies on a copper twisted pair infrastructure (the physical layer), organized into a circuit-switched architecture, with 64 Kbps channels set aside for each voice signal (data link, network, and transport layers). Applications such as flash hook signaling rely on elements ranging from the bottom (physical) to the top (applications) layers.[34]

The *Computer II* model in fact contemplated that enhanced services would be constructed in a layered manner, but it relied on an entirely different set of layers: Its fundamental assumption was that POTS services were the foundation on which

---

33. The OSI model was developed by the International Standards Organization to provide a common design framework for communications networks. Although the specific protocols developed as part of the OSI model were not widely adopted (and particular implementations may not follow the model rigorously), the concepts underlying the model are dominant in the computing world. The National Research Council followed a similar approach in devising its open data network architecture: That conceptual model incorporates a *bearer service* layer (sitting on top of the *network technology substrate*), a *transport* layer, a *middleware* layer, and an *application* layer. See NATIONAL RESEARCH COUNCIL, REALIZING THE INFORMATION FUTURE 47–51 (1994).

34. I owe this point to Stagg Newman.

enhanced services were built. One created an enhanced service by taking POTS service (or a similar but higher bandwidth service provided by the telephone company), using that service to transmit data, and adding data processing (and thus enhanced functionality). The underlying POTS transport was subject to regulation; the enhanced service—which was "enhanced" in the most literal sense—was not. That made perfect sense in the world of *Computer II*, back in 1980, and for years to come: It was perfectly natural for government to seek to regulate the underlying transport (which was, after all, offered for the most part by regulated monopolies), while eschewing any control over the enhancements. The 1996 Act, as the FCC has interpreted it, carried forward the same model.[35]

That key assumption does not work, though, in the IP world. IP maintains a sharp separation between the various layers of the OSI model. Different components of the network are responsible for the physical infrastructure, the transport of the underlying bits (using IP), and the applications (or services) that ride on top. That means, though, that the foundational assumption of *Computer II*—that an enhanced service is a basic service "plus"—no longer works.

In the IP world, there are no vertically integrated service offerings such as POTS that can be seen as the "foundation" of more elaborate offerings. Rather, the only foundation of any service offering is the underlying IP transport. Notwithstanding that a particular IP-based service offering may transmit information transparently, it does not therefore play the same role as POTS in the conventional telephony world. The transparent service offering does not provide transport for other, more elaborate IP-based service offerings; rather, raw IP provides transport for all.

As applied to the IP world, the basic–enhanced distinction does not serve the goal of allowing government to regulate underlying transport while leaving the enhancements to the marketplace. Instead, it only creates the anomalous distinction that a service is subject to regulation if it offers little functionality, but free from regulation if it offers somewhat more. It creates the anomalous result that two services, each deemed information services when viewed in isolation, may combine in a distributed environment to form an end-to-end offering magically deemed telecommunications. If regulators wish to carry forward into the IP world *Computer II*'s goal of attaching regulatory obligations to underlying transport, they need to aim those obligations more precisely.

Funding Universal Service

In rethinking universal service support in the modern telecommunications world, how should we draw the line between regulated and unregulated services? We

---

35. As the 1996 Act put it, a firm offers information services "via telecommunications."
47 U.S.C. § 153(20).

might best achieve the goals of *Computer II* by revising the universal service payment obligation to associate it not with service provision, but with the physical facilities along which the information moves.[36] *Computer II*, after all, sought to impose regulatory obligations on the underlying transport, and it is the physical layer that is associated with underlying transport in the most fundamental sense. A payment obligation tied to the ownership of qualifying facilities could apply without regard to whether the information moving via those facilities was in digital or analog form, or was packet- or circuit-switched.

Such an approach would have a variety of advantages. We could avoid the problems associated with determining which providers were providing telecommunications, making them subject to the assessment, and which were providing information services, leaving them exempt. Facilities ownership would trigger the obligation without regard to the nature of the traffic moving over those facilities. Such a rule might be able to do what the *Computer II* distinction itself can no longer do: It might effectuate *Computer II's* goal of imposing regulatory obligations on underlying transport, without burdening the service components higher up the protocol stack. It would thus vindicate *Computer II's* still valid judgment that, in order not to retard innovation, we should not impose regulatory costs on the new, still unfolding functionalities made possible by the marriage of silicon and data transmission.

Such an approach would be esthetically appealing: To the extent that the high-cost fund is designed to support the availability of physical infrastructure universally throughout the nation, it makes a nice symmetry to impose the associated costs on physical infrastructure. More consequentially, the approach would be technology neutral. Based on the assumption that a bit is a bit is a bit, no matter how transmitted, it would address the concerns of those who fear that a shift of telephone traffic away from circuit-switched voice to packet-switched data will undermine the entire subsidy structure. This seems important: In the ultimate analysis, it is hard to justify a regulatory scheme that assigns different consequences to provision of the same transport using different technologies. Such a scheme leads providers to make technology choices on the basis of regulatory arbitrage, not on the basis of which technology is most efficient, powerful, or inexpensive in the particular context.[37]

A facilities-based approach might well increase the share of universal service obligations ultimately paid, as costs are passed from suppliers to customers, by consumers of Internet-based services. That should not, though, be a dispositive objection. Conventional and IP networks are merging, so that it will no longer work simply to seek to insulate IP networks from regulation.[38] Rather, the goal should be to find ways to recast existing regulation (where it should not simply be jetti-

---

36. *See Report to Congress on Universal Service*, 13 FCC Rcd at 11535–36.

soned for circuit-switched and packet-switched networks alike) to be technology neutral and IP friendly, to make sense in an increasingly packet-switched world.

The FCC would have to overcome considerable practical difficulties, though, before it could adopt such an approach. How would the agency determine the amount of the fee paid by facilities owners? The agency currently sets universal service assessments as a percentage of the revenues a firm receives from end users for telecommunications. If the agency tied the fact of the payment obligation to physical facilities ownership, then it could sensibly tie the amount of the assessment to revenues only by looking to that limited set of revenues corresponding to physical transmission. Yet typically, a telecommunications (or information service) provider provides its customers with a combination of physical infrastructure, transport, and associated features and services, not just physical facilities. The revenues it receives are for the combination. Where a provider itself owns transmission facilities (rather than purchasing raw transmission from a third party) and provides its customers with an integrated service, it is not clear how one can isolate that portion of its revenues that correspond to raw transmission alone.[39]

---

37. Would such an approach be consistent with the Communications Act? 47 U.S.C. § 254(d) mandates that "[e]very telecommunications carrier providing interstate telecommunications services" contribute to universal service mechanisms "on an equitable and non-discriminatory basis." A facilities-based approach would be vulnerable to the objection that it did not require "every" carrier to contribute. One might argue, though, that under this approach non-facilities-based carriers would contribute (albeit indirectly) through the prices they paid for transmission, since those prices would reflect the facilities-based carriers' direct payments. Indeed, if the statute were read to impose an inflexible requirement that all carriers contribute directly, the current approach would not comply, since it is only the provision of telecommunications to end users that triggers the payment obligation. A carrier that does not serve end users is not required to contribute today.

The approach described in text might also be vulnerable to the argument that the FCC has no authority to impose payment obligations on facilities-based information service providers. Here, though, the Report to Congress on Universal Service provides the answer: Such a firm should be deemed to be providing telecommunications to itself, and thus falls within the FCC's authority to require "[a]ny . . . provider of interstate telecommunications . . . to contribute to the preservation and advancement of universal service if the public interest so requires." 47 U.S.C. § 254(d); see Report to Congress on Universal Service, 13 FCC Rcd at 11534–35.

38. To illustrate that convergence, a working group of the European Telecommunications Standards Institute, developing standards for computer-to-phone IP telephony, is considering an approach under which E.164 telephone numbers with a special country code would correspond to Internet addresses.

39. See Kevin Werbach, How to Price a Bit, RELEASE 1.0 (June 1998), at 1 (the "cost to send a bit of data across the Internet...is surprisingly complex. Networks involve a mix of fixed and variable investments, and pricing requires assumptions about demand levels, competition and usage patterns.").

It might be possible to tie universal service obligations to a metric other than revenues—say, to impose a fee based on a firm's asset investment in physical facilities. One possibility might be to make the fee proportional to the raw bandwidth of a firm's transmission facilities. Increasingly, though, carriers are creating bandwidth through improved multiplexing techniques rather than laying new fiber. It would be undesirable if a firm's implementation of such techniques led to a massive jump in its universal service obligations; that might discourage desirable experimentation and capacity expansion. Nor would it always be clear, in the case of innovative technologies, how much bandwidth to associate with a given facility. Indeed, for some technologies (e.g., unlicensed wireless spread spectrum), the notion of the bandwidth associated with a facility seems essentially meaningless. Further, it might introduce distortions if a fee were imposed in connection with idle transmission capacity; yet an approach that requires measuring the amount of bandwidth in use seems troublesome in its own way.

Looking beyond the networks traditionally subject to Title II, the unanswered questions become even more significant and pressing. If the FCC is to implement a technology-neutral solution, there is no obvious policy reason why it should exempt cable TV operators from the assessments paid by other transmission facilities owners. Arguably, the agency should impose similar assessments on any use of frequency spectrum. Under a technology-neutral approach, all communications transmission facilities—not just those historically used to provide telecommunications—are potentially in play.

The current regulatory structure evades those questions, but it does not adequately answer them. It excludes transmission facilities not traditionally subject to Title II by operation of statutory definitions: Cable operators need not make payments to the Universal Service Fund, in connection with the provision of cable service, because cable service is not telecommunications. That definitional analysis, though, does not resolve policy issues. Is it sensible policy that transmission over telephone plant presumptively triggers one set of obligations, whereas transmission over cable plant triggers another, when increasingly the same services can be provided over each?[40] If not, then we need to move beyond the current definitions by seeking to identify the communications policy objectives that underlie existing Title II regulation; deciding which of those objectives remains valid in today's semicompetitive environment; and with regard to each of those, figuring out which definitional boundary best achieves the regulatory goal—and what regulatory costs we impose by drawing that line.

---

40. See generally Barbara Esbin, "Internet Over Cable: Defining the Future in Terms of the Past," OPP Working Paper No. 30 (August 1998).

IMPLICIT SUBSIDIES

The Universal Service Fund is not the only—or even the most important—federal subsidy mechanism. Telephone pricing is characterized by a tangle of implicit as well as explicit cross-subsidies, and the implicit subsidies are larger than the explicit ones.[41] The most important federal implicit subsidy mechanism is interstate access pricing. Access charges today are an opaque blend of forward-looking economic cost, historic costs, and subsidies intended to depress local rates.[42] The FCC is seeking to remove the subsidy element from access charges, and to drive those charges down to a level more nearly approximating economic cost.[43]

Where should the Internet, and IP networks generally, fit within the access charge structure? As in the universal service context, there is no compelling reason for access charge obligations, in the long run, to turn on the telecommunications–information service distinction at all. Whereas universal service payments are pure subsidy, access charges include a cost-recovery element. There is no compelling reason why, in the long run, information service providers should not pay appropriate charges tied to the costs they impose on the local exchange. Rather, the goal should be to move access charges toward cost for telecommunications and information service providers alike.

The Status Quo

Currently, information service providers do not pay access charges. That exemption should continue for now. As the FCC has explained, it would make little sense to require ISPs to pay interstate access charges as currently constituted:

> [T]he existing access charge system includes non-cost-based rates and inefficient rate structures. [There is] no reason to extend this regime to an additional class of users . . . . The mere fact that providers of information

---

41. See Federal–State Joint Board on Universal Service, Report and Order, 12 FCC Rcd 8776, 8784 (1997), *appeal pending sub nom.,* Texas Office of Public Utility Counsel v. FCC, No. 97–60421 (5th Cir.).

42. See *id.* Historic (sunk) costs are based on a carrier's embedded technology. The FCC's forward-looking cost methodology, on the other hand, assumes the most efficient technology for reasonably foreseeable capacity requirements, deployed in the carrier's existing network configuration. Access charges today look to both of those cost methodologies, with a generous helping of additional revenue so that local telephone companies can price residential service below the incremental cost to them of providing that service.

43. See Access Charge Reform, First Report and Order, 12 FCC Rcd 15982, 15986–87, 15915–16004 (1997) (*Access Charge Reform Order*), aff'd, Southwest Bell Tel. Co. v. FCC, No. 97–2618 (8th Cir. Aug. 19, 1998). The FCC intends the explicit universal service mechanisms instead to shoulder the entire subsidy burden.

services use incumbent [LEC] networks to receive calls from their customers does not mean that such providers should be subject to an interstate regulatory system designed for circuit-switched interexchange voice telephony.[44]

I suggested in the previous section of this chapter that attempts simply to insulate IP networks from regulation are doomed to fail. However, that is not to say that one should blindly extend old rules to IP networks, no matter how inefficient or ill-advised that regulation is. The FCC is currently seeking to remove implicit universal service subsidies from interstate access charges. Against that backdrop, it would not be sensible to extend those subsidies to a new class of users, imposing distortions and inefficiencies on IP networks.

Beyond the Status Quo

However, the current exemption is not the end of the story—access charges, after all, recover costs and generate subsidies. In the absence of access charges or some comparable payment, there is no mechanism to cause ISPs to pay any congestion costs they impose on the local exchange. Any such costs, rather, are assigned to the local jurisdiction and spread among all local ratepayers (whether they subscribe to Internet access service or not).

The extent to which ISPs impose costs on the local exchange is hotly debated. The FCC's Local Competition Order, though, estimated a cost of .2 to .4 cents ($.002 to $.004) per minute as a default proxy for the traffic-sensitive component of local switching.[45] This figure, small as it is, suggests the potential for a mismatch between prices and economic costs where ISPs receive huge numbers of calls over the public switched network, because typically such a call is free to both caller (paying flat residential rates) and ISP (under standard local business rates, paying a flat fee for incoming calls). The matter is not simple—local switching costs appear to be essentially congestion costs, and the interested parties fiercely dispute the degree to which Internet access in fact generates congestion on the local network. The associated costs may well be zero except during peak periods. However, it seems reasonable to assume that Internet access does impose some costs on the local exchange not reflected in the rates ISPs pay.[46]

The legitimacy of any mismatch between prices and economic costs in this area is usually debated in federalism terms. ISPs urge that the local lines they buy fall within the intrastate jurisdiction, so that it is up to state regulatory

44. Access Charge Reform (Notice of Proposed Rulemaking), 11 FCC Rcd 21354, 21480 (1996) (footnote omitted). The agency confirmed this tentative conclusion in its *Access Charge Reform Order*, 12 FCC Rcd at 16133.

45. See Implementation of Local Competition, 11 FCC Rcd 15499, 15905, 16024–26 (1996).

commissions to decide whether there is an impermissible disparity between prices and costs. They continue that, in receiving large numbers of incoming calls while making few outgoing calls, ISPs are no differently suited from a variety of other local businesses (e.g., pizza parlors), and should not be singled out for different treatment.

That argument seems unsatisfactory, though, on a variety of levels. To the extent that the bulk of the inefficiencies and subsidies that characterize conventional telephony are built into the intrastate pricing structure, we should be wary of too quick a finding that any IP-based service is properly regulated as part of that structure. Moreover, the federalism argument seems wrong: ISPs provide what is in predominant part an interstate information service.[47] Customers use ISP facilities to exchange traffic with e-mail correspondents, Usenet news participants, websites, file transfer protocol servers, and other persons or devices without regard to jurisdictional boundaries.[48] Indeed, the entire point of Internet access is to enable communication with persons and sites across the globe.

Dial-up Internet access, to be sure, is an information service under the definitions discussed earlier, not a telecommunications service like long-distance POTS. It is by no means clear, though, why that should be relevant to a charge designed to recover actual costs imposed on the local exchange.[49] Because the Internet traffic passing over the local phone lines connecting end users and ISPs is predominantly jurisdictionally interstate, federal policy should govern how the costs associated with that traffic are allocated.

The fact that ISPs need not pay all of the costs they incur may lead to concrete distortions. Specifically, the most efficient way to move bits from end us-

---

46. The question whether ISPs impose uncompensated costs on the local exchange is different from the question whether they impose uncompensated costs on LECs. In the *Access Charge Reform Order*, the FCC found insufficient evidence that LECs suffered losses by virtue of Internet use. It noted that the carriers received revenue not only from ISPs' connections to the local exchange, but also from consumers' purchases of second lines and ISPs' purchases of leased lines to provision their internal networks. Moreover, the popularity of the Internet generated revenue through subscriptions to incumbent LECs' own Internet access services. *Access Charge Reform Order*, 12 FCC Rcd at 16133–34. These considerations, though, suggest that uncompensated costs in one area are balanced by monopoly profits in another. They do not speak to whether the rates paid by ISPs are related to the costs they impose (much less to whether either the profits LECs earn or the costs they incur are passed on to the ratepaying public).

47. See GTE Telephone Operating Cos., CC Docket No. 98–79 (rel. Oct. 30, 1998), at ¶¶ 22–26.

48. That traffic is sometimes stored on ISP computers along the way, but that storage (in a Web cache, Usenet news feed, or mail queue) is simply an intermediate step in a larger journey. "[T]he Commission traditionally has determined the jurisdictional nature of communications by the end points of the communication…" *Id.* at ¶ 17.

ers to ISPs may well be over digital, packet-switched links that bypass the public switched telephone network entirely (or that use customers' local loops, but leave the network before hitting a telephone switch). Yet ISPs' freedom from access charges may motivate them to stay on the circuit-switched network even where that is the less efficient solution. If ISPs were required to pay the economic costs of their connections to the circuit-switched network, then competitive LECs would have incentives to offer, and ISPs to buy, more efficient packet-switched connections. Incumbent LECs might then roll out their own comparable services in response.

On the other hand, the scenario just sketched out assumes the existence of local competition (i.e., it assumes that some firm is in fact offering packet-switched access in competition with the incumbent LEC). In the absence of local competition, reforming the rates paid by ISPs accomplishes nothing except that ISPs pay higher prices and incumbent LECs keep the money, because the monopoly providers have little incentive to develop ways to move the Internet traffic off the circuit-switched network.

One answer, thus, might be to postpone the imposition of any new charges on ISPs until after local competition emerges. There is a chicken-and-egg problem, though: One of the most important factors affecting the willingness of LECs (competitive or incumbent) to roll out packet-switched connectivity for ISPs is ISPs' willingness to buy that connectivity—yet current regulation diminishes ISPs' incentive to do so. Another answer, thus, might be for the FCC to announce now that ISPs will be required to pay a federally tariffed charge for connectivity to the circuit-switched network, reflecting actual economic costs, on the emergence of local competition in the relevant market. This would encourage competitive LECs to roll out packet-switched services directed at ISPs, knowing that the imposition of the federally tariffed charge on circuit-switched connectivity would level the playing field and make those services more attractive.

---

49. When the FCC initially established the access charge system in 1983, it contemplated that both basic and enhanced service providers would pay access charges. The agency explicitly defined *access service* to include "services and facilities...provided for the origination or termination of any interstate or foreign enhanced service." MTS and WATS Market Structure, 93 F.C.C.2d 241, 344, *on reconsid.*, 97 F.C.C.2d 682 (1983) (*MTS and WATS Market Structure Reconsideration Order*), 97 F.C.C.2d 834, *aff'd in relevant part sub nom.* NARUC v. FCC, 737 F.2d 1095 (1984), *cert. denied*, 469 U.S. 1227 (1985). It noted that enhanced service providers were users of access service in that they "obtained local exchange service or facilities which are used...for the purpose of completing interstate calls." *MTS and WATS Market Structure Reconsideration Order*, 97 F.C.C.2d at 711. On reconsideration, the agency changed its course, concluding that enhanced service providers would experience severe rate shocks if they were to pay the same access charges as long-distance carriers. It had no doubts, though, that such enhanced service providers "employ exchange service for jurisdictionally interstate communications." *Id.* at 715.

As in the previous section of this chapter, though, a wide range of questions remain. Would such an approach be practical? Is it unnecessary, because end-user demand for DSL and other packet-switched services, and competition from ISPs affiliated with incumbent LECs, will drive ISPs to seek packet-switched connections in any event? How can the agency determine the actual economic costs imposed by ISP circuit-switched connections on the local network? How would it measure competition? (Doesn't it have enough problems implementing the 1996 Act's famously problematic directive[50] that the Bell Operating Companies may provide in-region long-distance services only after they open up their local markets to competition?)

CONCLUSION

The distinction between regulated telecommunications and unregulated information services is at the center of the 1996 Telecommunications Act. That distinction, though, is rooted in the conventional telephone network; it does not work in the IP world. We need to develop new ways of reconciling old telephone regulation with new IP networks. For example, regulators should consider associating universal service payment obligations not with the provision of telecommunications, but with the ownership of transmission facilities. Such a rule might effectuate the underlying goals of the telecommunications–information service distinction, as current law cannot. Ultimately, we will have to reshape the rules governing both old and new technology if we are to find a structure that works.

---

50. 47 U.S.C. § 271.

Author Index

317

Subject Index